STRIPPING VOLTAMMETRY
IN CHEMICAL ANALYSIS

Kh. Z. Brainina

STRIPPING VOLTAMMETRY IN CHEMICAL ANALYSIS

Translated from Russian by P. Shelnitz
Translation edited by Dr. D. Malament

A HALSTED PRESS BOOK

JOHN WILEY & SONS
New York · Toronto

ISRAEL PROGRAM FOR SCIENTIFIC TRANSLATIONS
Jerusalem · London

Sole distributors for the Western Hemisphere
HALSTED PRESS, a division of
JOHN WILEY & SONS, INC., NEW YORK

Library of Congress Cataloging in Publication Data

Braĭnina, Kh'ena Zalmanovna.
   Stripping voltammetry in chemical analysis.

   Tranlation of *Inversionnaĭa vol'tamperometriĭa
tverdykh faz.*
   "A Halsted Press book."
   Bibliography: p.
   1. Voltammetry. I. Title
QDII5.B6413   621.37'46   74-13974
ISBN 0-470-09590-3

Distributors for the U.K., Europe, Africa and the Middle East
JOHN WILEY & SONS LTD., CHICHESTER

Distributors for Japan, Southeast Asia and India
TOPPAN COMPANY LTD., TOKYO AND SINGAPORE

Distributed in the rest of the world by
KETER PUBLISHING HOUSE JERUSALEM LTD.

ISBN 0 7605 1377 0
IPST cat. no. 22096

This book is a translation from Russian of
INVERSIONNAYA VOL'TAMPEROMETRIYA
TVERDYKH FAZ
Izdatel'stvo "Khimiya"
Moscow, 1972

Printed and bound by Keterpress Enterprises, Jerusalem
Printed in Israel

CONTENTS

PREFACE

Electrochemical analysis is one of the rapidly developing branche
branches of chemistry. This interest, particularly in regard to
voltammetry and polarography, is due to the possibility of easily
obtaining a vast amount of information on the kinetics of electrode
processes and solution composition. High sensitivity is one of the
most valuable aspects of these methods and is attained either by
using alternating current, which allows an increase in the signal-to-
noise ratio with the use of appropriate instrument circuits or by
prior concentration of a specific component on the electrode or in
the electrode material, that is, by drawing a larger amount of a
component into the electrode reaction zone. When the component is
concentrated as an amalgam, the process is called amalgam polarography with accumulation /1, 2/ and in the case in which a solid
metal or chemical compound is deposited, the process is termed
anodic stripping voltammetry (film polarography with accumulation
/2/). The major source of information in anodic stripping voltammetry is found in the polarization curves for the electrochemical
conversion of the solid deposited on the electrode surface or introduced into the electrode material.

Three variants of stripping voltammetry have been fairly widely
described in the literature. They correspond to the following electrode reaction types:

1) discharge-ionization of metals on an indifferent electrode, i.e.,
anodic stripping voltammetry of metals;

2) electrochemical oxidation-reduction of variable-valence ions
with the formation of an insoluble compound on the surface of an
indifferent electrode and the dissolution of these compounds, i.e.,
stripping voltammetry of variable-valence ions;

3) the formation of insoluble compounds on the surface of an
electroactive electrode as a result of anodic polarization of the
electrode and reduction of these compounds, i.e., cathodic stripping
voltammetry of anions.

Also promising is the use of electrochemical reactions of solids
mechanically introduced into an indifferent electrode. This variant
clearly should be defined as bulk anodic stripping voltammetry.

Stripping voltammetry of solids allows electrochemical determination of anions and of elements which do not form amalgams, are

not reduced to the metallic state, or are more inert than mercury.
Study of the deposition and dissolution of small quantities has yielded
valuable information on the mechanism of electrode reactions and
the detailed behavior of heterogeneous systems in which one of the
phases is emerging or disappearing. This information is of practical
importance, for example, in the use of electrochemical cells as
elements of electronic circuits in which deposition and dissolution
of thin layers of metals and compounds may be utilized to integrate
low currents.

The development of anodic stripping voltammetry has sparked
research aimed at finding an inert solid electrode which is stable to
electrochemical oxidation and reduction reactions as well as giving
an unusually high hydrogen and oxygen overvoltage and a low residual current. These requirements are largely fulfilled by specially
prepared graphite electrodes which are used widely today in stripping voltammetry of metals and variable-valence ions. These
electrodes may also be successfully employed in the polarography
of inorganic and organic substances, especially for carrying out
diverse electrochemical oxidation reactions.

Several general topics in the theory of electrode processes are
elucidated in the first chapter. The three types of stripping voltammetry are subjects of separate chapters (Chapters II—IV). These
chapters begin with a theoretical introduction. Also presented are
the uses of each method in the study of the kinetics of electrode
processes, the electrochemical characteristics of deposit-ion systems in solution and methods to determine the concentrations of a
given component. Experimental technique and electrode preparation
are described in Chapter V. Chapter VI deals with certain aspects
which have not yet been adequately treated, and the prospects for
stripping voltammetry. All the mathematical transformations have
been placed in the Appendix, which also includes comprehensive
tables containing information on the electrochemical properties of
elements and the means for determining these properties.

The author is grateful to Academician A. N. Frumkin for his help
in elaborating the theory of stripping voltammetry, Academician I. P.
Alimarin for useful recommendations in the analytic part of the book,
Academician Yu. S. Lyalikov, Professors A. I. Busev and A. G. Stromberg for constant help and constructive criticism of the work, B. I.
Khaikin for valuable advice in discussing the results, her coworkers, E. M. Roizenblat, V. B. Belyavskaya, T. A. Krapivkina, N. K.
Kiva, and E. Ya. Sapozhnikova for participation in carrying out the
experiments described, and E. Ya. Neiman for help in preparing the
manuscript.

The author will always remember with gratitude the constant
interest and help rendered by the late Dr. S. I. Sinyakova.

The Author

LIST OF SYMBOLS

a, a_{∞} — activity of a precipitate on an electrode and activity of the corresponding macro-phase /3/, mol/cm^3;

$$a' = \frac{a}{a_{\infty}}$$

a'_{∞} = 1 mol/cm^3 (introduced to maintain the correct dimensions);

c^0 — concentration of ions in the solution bulk, g·ion/cm^3;

c^0_{min} — minimum bulk concentration of ions at which formation of a precipitate of the compound on the electrode is possible, g·ion/cm^3;

$c^s, c(0)$ — concentration of ions close to the electrode surface, mol/cm^3;

D — diffusion coefficient, cm^2/sec;

E — energy of activation of an electrode process, J/mol;

F — Faraday's number, C/g·eq;

f — activity coefficient of a substance;

i — current strength, A;

i_- — cathodic component of the current, A;

i_+ — anodic component of the current, A;

i_{dif} — limiting diffusion current, A;

i_{max} — maximum current value, A;

K — thermodynamic equilibrium constant of a chemical reaction;

K' — formal equilibrium constant of a chemical reaction;

k_s — rate constant of an electrode process at standard potential, cm/sec;

k_- and k_+ — rate constants of cathodic and anodic components of an electrochemical reaction, cm/sec;

k_1, k_2 — rate constants of forward and reverse chemical reactions, sec^{-1};

l — thickness of deposit film, cm;

n — number of electrons;

Q — quantity of electric charge equivalent to the quantity of deposited metal (compound), C;

R, O — reduced and oxidized forms of an electroactive substance;

S — area of electrode, cm^2;

t — time, sec;

v — voltage scanning rate, V/sec;

α, β — transfer coefficients of electrode process;

γ — proportionality coefficient /3/, k^{-1};

δ — thickness of diffusion layer, cm;

μ — thickness of kinetic layer, cm;

τ_1 — deposition time, sec;

$\tau, \varkappa, u, j$ — dimensionless parameters, equivalent to time, length, concentration, and current strength;

φ — electrode potential, V;

φ_1 — initial potential value, V;

φ^0 — standard potential, V;

φ_e — equilibrium potential, V;

$\varphi_{1/2a}$ or $\varphi_{1/2c}$ — half-wave potentials of anodic or cathodic polarograms, V.

PERIODIC TABLE OF THE CHEMICAL ELEMENTS AND SOME POSSIBILITIES
OF STRIPPING VOLTAMMETRY OF THE SOLID PHASES

Period	Series	Group of elements								
		I	II	III	IV	V	VI	VII	VIII	
I	1	H								He
II	2	Li	Be	B	C	N	O	F		Ne
III	3	Na	Mg	Al	Si	P	S	Cl		Ar
IV	4	K	Ca	Sc	Ti	V	Cr	Mn	Fe Co Ni	
	5	Cu	Zn	Ga	Ge	As	Se	Br		Kr
V	6	Rb	Sr	Y	Zr	Nb	Mo	Tc	Ru Rh Pd	
	7	Ag	Cd	In	Sn	Sb	Te	I		Xe
VI	8	Cs	Ba	La*	Hf	Ta	W	Re	Os Ir Pt	
	9	Au	Hg	Tl	Pb	Bi	Po	At		Rn
VII	10	Fr	Ra	Ac**	Ku					

*Lanthanides	Ce	Pr	Nd	Pm	Sm	Eu	Gd	Tb	Dy	Ho	Er	Tm	Yb	Lu
** Actinides	Th	Pa	U	Np	Pu	Am	Cm	Bk	Cf	Es	Fm	Md	No	Lr

Elements whose solid-phase reactions and methods of determination are described

Elements which will probably be used in the future in solid-phase voltammetry

Chapter I

THE THEORY OF ELECTRODE PROCESSES

An electrode process consists of a number of consecutive steps /4—18/.

1. Transport of the reactant from the bulk of the solution to the reaction zone.

2. A chemical reaction in which an electroactive product is generated. This product may reach the electrode-solution interface by diffusion, convection, migration, or adsorption on the electrode surface. Such a chemical reaction is usually called a preliminary reaction.

3. An electrochemical reaction, an electron transfer or Vetter transition reaction step /4/.

4. Removal of the electrochemical reaction products from the reaction zone. This may be achieved by diffusion, convection, and migration when the electrochemical reaction products are soluble and do not interact chemically with components of the solution.

If the reaction products are insoluble or interact chemically with components of the solution, we must consider the processes associated with removal of the chemical interaction products into the solution bulk or accumulation of these products on the electrode-solution phase boundary. Chemical reactions following transfer reactions are called subsequent reactions.

Let us consider two examples.

Electrochemical evolution of hydrogen. This process consists of the following steps /4, 5, 19/: transport of hydrogen ions to the electrode surface, a reduction reaction accompanied by recombination of adsorbed hydrogen atoms, and removal of the reaction product from the electrode surface:

$$2H^+ + 2e \longrightarrow 2H_{ads} \qquad 2H_{ads} \longrightarrow H_2$$

Depending on the experimental conditions and the electrode material, the rate-limiting step is the reduction reaction (delayed discharge-ionization) /5/ or recombination of atoms /4, 5/.

Oxidation and reduction of cadmium in a solution containing CN^- ions. This process /20/ entails ion transport to the electrode surface, a chemical reaction generating electroactive ions, an electrochemical reaction accompanied by a subsequent chemical conversion or removal of the reaction product into the solution bulk by diffusion:

$$\begin{rcases} \text{Cd} + 2\text{CN}^- \longrightarrow \text{Cd(CN)}_2^- + e \\ \text{Cd(CN)}_2^- + \text{CN}^- \longrightarrow \text{Cd(CN)}_3^{2-} \\ \text{Cd(CN)}_3^{2-} \longrightarrow \text{Cd(CN)}_3^- + e \\ \text{Cd(CN)}_3^- + \text{CN}^- \rightleftharpoons \text{Cd(CN)}_4^{2-} \end{rcases} \text{oxidation}$$

$$\begin{rcases} \text{Cd(CN)}_4^{2-} \rightleftharpoons \text{Cd(CN)}_3^- + \text{CN}^- \\ \text{Cd(CN)}_3^- + 2e \rightleftharpoons \text{Cd} + 3\text{CN}^- \end{rcases} \text{reduction}$$

The rate of a complicated process is determined by the rate of the most difficult step (also sometimes called the limiting or slow step).

In electrochemical kinetics, the electron-transfer process is studied in greatest detail, since this step determines the nature of the current-potential dependence.

We make the following assumptions.

Firstly, we shall assume that there is an indifferent electrolyte in the solution, the concentration of which is sufficient so as to allow us to neglect the ion migration and the potential drop in the diffuse portion of the double layer.

Secondly, since one of the steps is delayed, we shall assume that all the remaining steps reach an equilibrium state instantaneously. Processes not complicated by the formation of a new phase and processes resulting in the emergence or disappearance of a new solid phase will be considered separately.

The effect of the specific adsorption of surface-active substances on the double layer structure will not be treated.*

PROCESSES NOT COMPLICATED BY THE FORMATION OF A NEW PHASE

ELECTROCHEMICAL POLARIZATION

Here the rate-limiting step of the process is electron transfer. We shall call such a process irreversible and talk about the electro-chemical polarization of the electrode.

The theory of these irreversible processes, which has been termed delayed discharge-ionization theory, was developed for hydrogen discharge-ionization /5/ and may be applied to other cases as shown by subsequent investigations /4/.

---

* This problem is the subject of a special monograph by Delahay /21/.

In general, an electrochemical reaction may be represented by the equation:

$$O + ne \rightleftarrows R \tag{I.1}$$

When the reaction proceeds to the right, the flow of electrons is from the electrode to the solution. This process corresponds to the cathodic current component (i_-). When reaction (I. 1) proceeds to the left, the flow of electrons is from the solution to the electrode, and the process corresponds to the anodic current component (i_+). The current (i) flowing through the electrode-solution interface is the algebraic sum of these quantities. The sign and value of i determine the direction of the reaction and the amount of material which undergoes the electrochemical conversion.

The fairly high activation energy of electron transfer is a factor which hinders electrochemical reactions. The activation energy of cathodic and anodic processes depends on the electrode potential, more specifically on the potential difference arising in the dense portion of the electric double layer /4—6, 8/:

$$E_- = E_-^0 + \alpha nF\varphi \tag{I.2}$$

$$E_+ = E_+^0 - (1 - \alpha)nF\varphi \tag{I.3}$$

where E_-^0 and E_+^0 are the values of activation energy at $\varphi = 0$.

The rate of electron transfer in a given direction, such as from the reducing agent R to the electrode, is proportional to the activity of the reducing agent in the layer adjacent to the electrode and the exponential factor in which the activation energy is included. Thus,

$$i_+ = nFSk'_+ a_R \exp\left(-\frac{E_+}{RT}\right) = nFSk_+ a_R \exp\left[\frac{(1 - \alpha)nF\varphi}{RT}\right]^* \tag{I.4}$$

$$k'_+ = k_+ \exp\left(-\frac{E_+^0}{RT}\right)$$

Likewise, for the cathodic current component we may write

$$i_- = nFSk_- a_O \exp\left(-\frac{\alpha nF\varphi}{RT}\right)^* \tag{I.5}$$

The total current equals

$$i = nFS\left\{k_+ a_R \exp\left[\frac{(1 - \alpha)nF\varphi}{RT}\right] - k_- a_O \exp\left(-\frac{\alpha nF\varphi}{RT}\right)\right\} \tag{I.6}$$

---

* Here a_R and a_O are the activities of the components R and O near the electrode surface, which are equal to the activities of these components in the solution bulk under the assumption adopted.

At the equilibrium potential between the oxidized and reduced forms of the electroactive substance, there is equilibrium and the current vanishes. Then

$$k_+ a_R \exp\left[\frac{(1-\alpha)nF\varphi_p}{RT}\right] = k_- a_O \exp\left(-\frac{\alpha nF\varphi_p}{RT}\right) \qquad (\text{I. }7)$$

Hence, φ_p equals:

$$\varphi_p = \frac{RT}{nF}\ln\frac{k_-}{k_+} + \frac{RT}{nF}\ln\frac{a_O}{a_R} \qquad (\text{I. }8)$$

This relation is simply the Nernst equation, in which the standard potential $\varphi^0 = (RT/nE)\ln(k_-/k_+)$ is expressed in terms of the forward and reverse electrochemical reaction rates. Equation (I. 6) may be represented as

$$i = nFSk_S\left\{a_R \exp\left[\frac{(1-\alpha)nF}{RT}(\varphi-\varphi^0)\right] - a_O \exp\left[-\frac{\alpha nF}{RT}(\varphi-\varphi^0)\right]\right\} \qquad (\text{I. }9)$$

In this expression, a single rate constant k_S appears instead of k_+ and k_-:

$$k_S = k_+ \exp\left[\frac{(1-\alpha)nF}{RT}\varphi^0\right] \qquad (\text{I. }10)$$

$$k_S = k_- \exp\left(-\frac{\alpha nF}{RT}\varphi^0\right) \qquad (\text{I. }11)$$

The value of k_S, in contrast to k_+ and k_-, is independent of the potential scale selected.

When the anodic electrode polarization of the electrode is fairly large, i_+ is much greater than i_-, and the value of the total current is determined by the anodic component

$$i = i_+ = nFSk_S a_R \exp\left[\frac{(1-\alpha)nF}{RT}(\varphi-\varphi^0)\right] \qquad (\text{I. }12)$$

Likewise, when the cathodic polarization is adequate,

$$i = i_- = nFSk_S a_O \exp\left[-\frac{\alpha nF}{RT}(\varphi-\varphi^0)\right] \qquad (\text{I. }13)$$

Taking the logarithms of these expressions, we obtain equations similar to Tafel's equation:

$$\varphi = \varphi^0 - \frac{RT}{(1-\alpha)nF} \ln nFSk_S - \frac{RT}{(1-\alpha)nF} \ln a_R + \frac{RT}{(1-\alpha)nF} \ln i \qquad (\text{I. }14)$$

$$\varphi = \varphi^0 + \frac{RT}{\alpha nF} \ln nFSk_S + \frac{RT}{\alpha nF} \ln a_O - \frac{RT}{\alpha nF} \ln |i| \qquad (\text{I. }15)$$

Using empirical dependences between the current and potential of an electrode, we easily find the values of α and k_S characterizing the electrochemical reaction kinetics from these equations.

CONCENTRATION POLARIZATION

Concentration polarization is caused by a concentration change in the layer adjacent to the electrode owing to the diminished supply or removal of the electroactive substance and delayed chemical reactions in the solution generating the electroactive substance. Let us consider the case in which the rate-limiting step is supply or removal of the electroactive substance to or from the electrode surface. We assume that all chemical processes and electron transfer reactions are at equilibrium. In this case, the electrode potential may be calculated from the Nernst equation, using the activity of the oxidized and reduced forms of the component participating in the reaction near the electrode surface rather than in the solution bulk. The component concentrations (and their activities) in the layer adjacent to the electrode are functions of the current (i) and the time (t). The expression for the electrode potential may be written as:

$$\varphi = \varphi^0 + \frac{RT}{nF} \ln \frac{a_O(i,t)}{a_R(i,t)} \qquad (\text{I. }16)$$

The transfer of a component from the solution bulk to and from the electrode surface occurs by diffusion when the reaction product is soluble. Since the change in the concentration of the electroactive substance is small compared with the total electrolyte concentration, the change in the activity coefficient (f) may be neglected, and concentrations may be used instead of activities. A concentration gradient usually depends on time. The relationship between these quantities is given by Fick's second law:

$$\frac{\partial c_j}{\partial t} = D \frac{\partial^2 c_j}{\partial x^2} \qquad (\text{I. }17)$$

The concentration of component c_j is a function of x (the distance from the electrode surface) and t. To find c_j, we must establish initial and boundary conditions corresponding to the specific electrolysis conditions.

Let us consider, for example, the reaction $O + ne \rightleftarrows R$ occurring on a flat electrode in a cell whose length in the direction perpendicular to the electrode surface is sufficiently large to be assumed infinite. The solution is not stirred. The supply of component O to the electrode and the removal of R to the solution bulk occur by diffusion. The the following conditions are fulfilled on the metal-solution interface $(x=0)$:

$$\frac{c_0(0,t)}{c_R(0,t)} = \frac{f_R}{f_O} \exp\left[\frac{nF}{RT}(\varphi - \varphi^0)\right] = \theta \tag{I.18}$$

$$D_0 \left.\frac{\partial c_O}{\partial x}\right|_{x=0} = -D_R \left.\frac{\partial c_R}{\partial x}\right|_{x=0} \tag{I.19}$$

Equation (I.18) follows directly from equation (I.16), and equation (I.19) describes diffusion as a continuous process.

At the starting moment ($t=0$), component O is dispersed uniformly in the solution: $c_O(x, 0) = c^0$; $c_R(x, 0) = 0$ (c^0 is the concentration in the solution bulk). In addition, as $x \longrightarrow \infty$, $c_O \longrightarrow c^0$ and $c_R \longrightarrow 0$ and the current is proportional to

$$i = nFSD \left.\frac{\partial c}{\partial x}\right|_{x=0} \tag{I.20}$$

Solution of equations* (I.17)–(I.20) relates the current, time, and potential /8/:

$$i = nFSD_O^{1/2}c^0 \frac{1}{\pi^{1/2} t^{1/2}(1 + \varepsilon\theta)} \tag{I.21}$$

$$\text{where } \varepsilon = \left(\frac{D_O}{D_R}\right)^{1/2}.$$

When θ approaches zero and $\varphi \longrightarrow -\infty$, the current becomes equal to the limiting diffusion current

$$i = i_{\text{dif}} = nFSD_O^{1/2}c^0 \frac{1}{\pi^{1/2} t^{1/2}} \tag{I.22}$$

Taking relations (I.18), (I.21), and (I.22) into consideration, we may obtain the following expressions for the electrode potential:

$$\varphi = \varphi^0 - \frac{RT}{nF} \ln\left(\frac{f_R}{f_O}\right)\left(\frac{D_O}{D_R}\right)^{1/2} + \frac{RT}{nF} \ln \frac{i_{\text{dif}} - i}{i} \tag{I.23}$$

*The methods for solving partial differential equations are treated in specialized books /22/.

or

$$\varphi = \varphi_{1/2} + \frac{RT}{nF} \ln \frac{i_{\text{dif}} - i}{i} \qquad (\text{I}.24)$$

where $\varphi_{1/2} = \varphi^0 - \dfrac{RT}{nF} \ln \left(\dfrac{f_R}{f_O}\right)\left(\dfrac{D_O}{D_R}\right) -$ is the potential corresponding

to the current equal to half of the limiting current, i.e., the value of the potential at the point of inflection in the current-potential curve.

We note that in the electrolysis of a non-stirred solution the current, including the limiting current, is time-dependent.

Under steady-state hydrodynamic conditions (rotating-disk electrode or stirred solution), the concentration gradient of the electroactive substance arises only in a layer of finite thickness (δ), termed the diffusion layer /4, 5, 23/, and the concentration is independent of the time. In this case, we may assume that the concentration gradient is constant in the entire layer.

$$\frac{\partial c_O}{\partial x} = \frac{c^0 - c(0)}{\delta} \qquad (\text{I}.25)$$

The current is determined by the formulas

$$i = nFSD \frac{c^0 - c(0)}{\delta} \qquad (\text{I}.26)$$

$$i_{\lim} = i_{\text{dif}} = nFSD \frac{c^0}{\delta} \qquad (\text{I}.27)$$

The limiting diffusion current does not vary with time. This current is directly proportional to the concentration of the electroactive substance in the solution and depends on the stirring intensity since this intensity determines the thickness of the diffusion layer /4/.

The rate-limiting step of an electrode process may be the chemical reactions occurring at the electrode surface. Such reactions either precede the electrochemical reaction and generate the electroactive substance or follow the reaction and use the products of the electrochemical conversion. Chemical reactions may also occur in parallel with electron transfer. A current dependent on the rate constants of chemical reactions is called a kinetic current. Electrode processes with chemical steps have been described in detail by Mairanovskii /6/. There are two methods of investigating such processes, namely, the approximate but direct method of Brdička and Wisner /24, 25/ based on the concept of a reaction layer, and an exact method which entails the solution of partial differential equations with initial and boundary conditions corresponding to the specific processes under study.

Let us consider the simplest electrode process with a preliminary unimolecular chemical reaction

$$A \underset{k_2}{\overset{k_1}{\rightleftharpoons}} B \overset{e}{\longrightarrow} C \tag{I.28}$$

The electroactive substance B is in equilibrium with component A which is not a depolarizer. The equilibrium is defined by the constant $K = c_A^0/c_B^0$, and the rate of establishing the equilibrium depends on the forward (k_1) and reverse $(k_2 = k_1 K)$ chemical reaction rate constants. If $K \gg 1$, the equilibrium is shifted toward component A, and then the diffusion current component is small owing to the low concentration of B in the solution. In this case, the current will be caused primarily by the flow of the substance arising as a result of the chemical reaction.

When the current flows, the equilibrium shift is obviously maximized near the electrode surface and decreases with increasing distance into the bulk of the solution. The layer of thickness μ in which the equilibrium shift is observed is called the reaction layer, the volume of which equals μS.

The limiting kinetic current is described by the relationship

$$i_{\lim} = nF\mu S k_1 c_A(\mu) \tag{I.29}$$

where $c_A(\mu)$ is the concentration of the electrochemically inactive component A in the reaction layer.

The value of $c_A(\mu)$ is determined by the rate of the conversion $A \longrightarrow B$. As the rate of this reaction increases, $c_A(\mu)$ decreases and may vanish entirely. In that case, the kinetic current is caused by the diffusion of A from the solution.

The thickness of the reaction layer is determined /26/ by the length of the path through which the electrochemically active particle travels during its average lifetime. The average lifetime of a particle is inversely proportional to the rate of the reaction resulting in its destruction /27—29/. Hence,

$$\mu = \sqrt{\frac{D}{k_1 K}} \tag{I.30}$$

Substituting this value for μ into equation (I.29), we obtain the following expression for the limiting current:

$$i_{\lim} = nFS D^{1/2} k_1^{1/2} K^{-1/2} c_A^0 \tag{I.31}$$

Hanuš /30/ proposed a somewhat different approach. He assumed that the concentration gradient of the electroactive substance B within the reaction layer is a constant and is determined by the relationship

$$\frac{\partial c_B}{\partial x}\bigg|_{x=0} = \frac{c_B(\mu) - c_B(0)}{\mu} \tag{I.32}$$

where $c_B(\mu) = c_B^0$ and $c_B(0)$ are the concentrations at the boundaries of the kinetic layer.

The kinetic current equals

$$i_{\lim} = nFSD\left(\frac{\partial c_B}{\partial x}\right)_{x=0} \tag{I.33}$$

Substituting equations (I.32) and (I.30) into equation (I.33) and recalling that $K = c_A^0/c_B^0$, we easily show that when $c_B(0)$ equals zero, equations (I.31) and (I.33) are identical.

The second method mentioned above, which is restricted to the solution of differential equations with specific initial and boundary conditions, is considerably more accurate than the method based on the concept of a reaction layer. In this case, the thickness of the kinetic layer μ is not a theoretical parameter, and is used only for evaluations. The value of μ may be found here from functions describing the distribution of the reactant concentrations in the solution layer adjacent to the electrode. This method is applicable to electrode processes complicated by various chemical reactions preceding electron transfer /11/, by regeneration of the depolarizer /10/, by processes with a slow electrochemical step and fast preliminary and subsequent reactions /17/, etc. /12—15/.

Let us consider, as an example, the above-cited case of the reduction of substance B which occurs according to the scheme represented by (I.28). Substance B is reduced on the electrode instantaneously, while substance A does not undergo any electrochemical conversions. Here the distribution of the concentrations c_A and c_B is described by the equations:

$$\frac{\partial c_A(x,t)}{\partial t} = D_A\frac{\partial^2 c_A(x,t)}{\partial x^2} + k_2 c_B(x,t) - k_1 c_A(x,t) \tag{I.34}$$

and

$$\frac{\partial c_B(x,t)}{\partial t} = D_B\frac{\partial^2 c_B(x,t)}{\partial x^2} + k_1 c_A(x,t) - k_2 c_B(x,t) \tag{I.35}$$

with the following initial and boundary conditions:

$$\text{or} \quad \left.\begin{array}{l} t=0 \\[4pt] x \longrightarrow \infty \end{array}\right\} \quad \frac{c_A}{c_B}=K \qquad c_A + c_B = c^0$$

$$\left.\begin{array}{l} t>0 \\[4pt] x=0 \end{array}\right\} \quad c_B(0,t)=0 \qquad D_A \left.\frac{\partial c_A}{\partial x}\right|_{x=0}=0$$

Koutecký and Brdička /27/ solved this problem by assuming that $D_A = D_B = D$ and that the solution equilibrium is shifted toward **A**, i.e. $K \gg 1$ and $c^0 = c_A$. These authors obtained the following expression for the current[*];

$$i = nFSc^0 D^{1/2} k_1^{1/2} K^{-1/2} \exp(\lambda^2) erfc\lambda \tag{I. 36}$$

where $\lambda = (k_1 K^{-1} t)^{1/2}$, and $erfc$ is an error function addition.

In the case of a slow chemical reaction ($k_1 t$ is small and $\lambda \ll 1$), the equilibrium concentration distribution of **A** is undisturbed, and the current is determined by the rate constant of the preliminary chemical reaction and is a kinetic current. If $\lambda < 0.1$, $\exp(\lambda^2) erfc\lambda \longrightarrow$

$$i_{\lim} = nFSD^{1/2} k_1^{1/2} K^{-1/2} c_A^0 \tag{I. 37}$$

This equation is identical to equation (I. 31). Thus, a characteristic feature of kinetic currents is their dependence on the rate and equilibrium constants of the chemical reactions. In the other limiting case ($\lambda \gg 1$) which corresponds to a fast chemical reaction, the current is determined by the diffusion of **A** from the solution bulk and equals the limiting diffusion current /8/.

Apart from the cases just cited, the electrode process may be complicated by additional steps of crystallization, catalytic process and the adsorption of electroactive and foreign substances.

The equation relating the current and potential of an electrode must contain all the parameters of the process; namely, the rate constant of the electrode process, the rate constants of all possible chemical reactions, and the diffusion coefficients of all substances involved in the reaction. Unfortunately, it is impossible mathematically to describe such processes in a general form. This problem usually solved for more or less particular cases, and by attaining limiting conditions, criteria are found by which we can judge the rate limiting step of the process.

We have already proven that each case has characteristic features which make it possible to isolate it in an experimental study of the kinetics of an electrochemical conversion. Thus, for example, the presence of a limiting current on the polarization curve indicates

---

[*] The symbols of Koutecký and Brdička have been replaced by those adopted in this book.

concentration polarization caused by slow diffusion or a chemical reaction. These processes are distinguished by the fact that a limiting diffusion current depends on the hydrodynamic conditions while a limiting kinetic current does not. The latter has a higher temperature coefficient since the rate constants of chemical reactions vary more strongly with the temperature than do diffusion coefficients. If the rate-limiting step is electron transfer, the current does not reach a limiting value but rather depends on the electrode potential.

Isolation of any rate-limiting step, i.e., the creation of experimental conditions such that all the other steps will be at equilibrium, enables us to find the corresponding kinetic constants seen in the equations presented.

PROCESSES COMPLICATED BY THE FORMATION AND DISSOLUTION OF SOLID PHASES

Many electrode processes involve the formation or dissolution of solid phases such as the discharge or ionization of metals on indifferent electrodes, deposition and dissolution of oxides and salts on the electrode-solution interface, and electrochemical conversions of one solid into another. These processes are the basis of electroplating, galvanoplastics, and electrometallurgy and are used in chemical current sources as well as in analytical chemistry.

The laws governing electrode processes described in the preceding section are in general valid for processes involving the formation or disappearance of a solid phase. Here we must consider the change in the electrode surface and the nature and properties of the new phase /31—33/. The change in the electrode surface is usually neglected and we assume that the activity of the deposit formed at once becomes equal to the activity of the corresponding solid phase. Then the system obeys the Nernst equation in the usual abridged form and the boundary condition (I. 18) for equation (I. 17) is written very simply: $c_0(0,t) = 0$.

However, such an assumption is incorrect for incompletely coated electrodes. This has been found in an experimental study of electrolytic deposition of metals from very dilute solutions /34—48/.

Studying electrolytic deposition processes of small amounts of metals, Rogers and Stehney /37/ proved that a modified Nernst equation may be derived, which gives with some accuracy the relationship between the electrode potential and several experimental parameters. The authors started from the fact that the activity of a metal on an inert electrode is determined by the product of the electrode portion

occupied by the deposit and the activity coefficient of the deposited
metal. They also assumed that the electrode is coated initially by a
monolayer and that a second layer does not form until the surface is
completely covered by the first layer of atoms. In this case, the
activity of the deposited material is described by the expression

$$a = \frac{N \cdot S_a \cdot f_{Me}}{S} \tag{I.38}$$

where N is the number of atoms deposited, S_a the area occupied by
one atom in cm^2, and f_{Me} is the activity coefficient of the metal.

The number of atoms deposited on the electrode before equilibrium is attained equals

$$N = (c^0 - c)V \cdot N_A \tag{I.39}$$

where $c^0 - c$ is the difference between the initial and equilibrium
concentrations of the ions of the metal being discharged, V is the
volume of the solution, and N_A is Avogadro's number.

The equilibrium potential of an incompletely coated electrode is
described by the relationship /49/:

$$\varphi_e = \varphi^0 + \frac{RT}{nF} \ln \frac{cSf_{Me^{n+}}}{(c^0 - c)VN_A S_a f_{Me}} \tag{I.40}$$

or

$$\varphi_e = \varphi_{50\%} - \frac{RT}{nF} \ln \frac{c^0 - c}{c} \tag{I.41}$$

where

$$\varphi_{50\%} = \varphi^0 + \frac{RT}{nF} \ln \frac{Sf_{Me^{n+}}}{VN_A S_a f_{Me}}$$

where $\varphi_{50\%}$ is the potential at which 50% of the metal ions present
in the solution are deposited.

This shows that the potential $\varphi_{50\%}$ is independent of the initial
concentration of the electroactive ions in the solution but rather is
determined by the electrode surface, the solution volume, and the
activity coefficient of the metal deposit. The resulting equation
describes the electrolytic deposition of a metal from a very dilute
solution on an incompletely coated electrode. Total recovery of the
metal is possible by a potential shift in a negative direction.
According to our assumptions, after the surface is coated with a
monolayer of deposit, the activity of the layer becomes equal to

that of the corresponding solid phase and the system obeys the
Nernst equation in the usual form. Rogers and Stehney showed that
similar reasoning holds true for the formation of oxide films.

Byrne, Rogers, and Griess /43/ obtained experimental curves of
the amount of precipitated silver (in %) as a function of the electrode
potential for a concentration range including both deposition of
traces and of fairly large amounts of metal. In the first case, the
curves fit equation (I. 41). The insignificant shift in the value of $\varphi_{50\%}$
in the negative direction observed in the experiments when the concentration of silver ions in the solution was increased differs from
the case corresponding to a completely coated electrode whose
potential shift must have the opposite sign according to the Nernst
equation. The transition from the deposition of traces to the deposition of quantities of silver adequate for the formation of one or
several atomic layers is accompanied by the appearance on the
curve of a transition region from the more positive to the less positive potentials, where the amount of metal deposited is independent
of the electrode potential.

Byrne and Rogers /44/ compared the dimensions of gold, palladium, and platinum electrodes with the area occupied by a monoatomic
layer of silver and proved that the first value exceeds the second
when the system obeys equation (I. 41). The opposite relation is
observed when the system may be described using the abridged
Nernst equation and the areas are commensurate in the intermediate
case. According to these authors, silver forms a monolayer on
platinum, palladium, and gold during deposition. In the electrochemical deposition of bismuth and lead on silver and gold /50—53/ electrodes, a monolayer deposit arises on the atoms deposited first. In
this case, the system is described by the abridged Nernst equation
even when the electrodes are partially coated by deposit which is
contrary to Rogers and Stehney's concepts. Mercury, iron, and cadmium are also deposited on the active centers of a monocrystalline
platinum electrode surface followed by growth of the nuclei formed
/54/.

These examples, as well as the comprehensive investigations of
Ziv et al. /45—48/, lead to the conclusion that the state of a deposited
metal and its distribution on the surface of an indifferent electrode
depend on such difficult factors as the nature and method of preparing
the electrode surface, interaction forces between atoms of the same
and different metals, adsorption phenomena on the electrode-solution
interface, and the amount of material deposited. These factors have
not yet yielded to mathematical analysis. Therefore, it is still
apparently impossible to give an accurate determination of the
activity of a deposit on an indifferent electrode. It is thus convenient
to use the concept of activity as a variable characterizing the

difference between the properties of small amounts of deposit and the properties of the respective macrophase /3, 55/. The value of the activity (a) must be a function of Q and, with sufficiently high Q, becomes equal to the activity of the corresponding pure phase, whic' we denote by a_∞ (a_∞ expressed in mole fractions equals 1). When Q is small, the deposit activity /37/ must be proportional to Q. One of the possible interpolation formulas satisfying these two condition is

$$a = a_\infty \left[1 - \exp\left(-\gamma Q\right)\right] \tag{I.42}$$

where γ is a parameter characterizing the properties of an electrode-deposit system.

In fact, at a sufficiently low value of Q $(\gamma Q \ll 1)$, the exponential may be expanded in series, and we may limit ourselves to the first two terms of the expansion. Then

$$a = a_\infty \gamma Q \tag{I.43}$$

The activity of the deposit is directly proportional to the amount of deposit on the electrode. We shall call such deposits microphase

Conversely, when Q is fairly high $(\gamma Q \gg 1)$, the exponential term is negligible and the activity of the deposit ceases to depend on Q $(a = a_\infty)$. We shall call such deposits macrophases.

The proposed formula may be checked by measuring the equilibrium potentials of electrodes covered by different amounts of depos in solutions containing potential-determining ions.

Under steady-state hydrodynamic electrolysis conditions, Q is directly proportional to the concentration of the electroactive ions (c^0) and the deposition time (τ_1):

$$Q = k_{\text{dif}} c^0 \tau_1 \tag{I.44}$$

where k_{dif} is a proportionality coefficient.

Considering equation (I.42), we may represent the expression for the equilibrium as

$$\varphi_e = \varphi_1^0 \pm \frac{RT}{nF} \ln \frac{c^0 f}{a_\infty \left[1 - \exp(-\gamma Q)\right]} \tag{I.45}$$

where

$$\varphi_1^0 = \varphi^0 + \frac{RT}{nF} \ln a_\infty$$

In the limiting cases treated above, this expression transforms into the following two equations:

$$\gamma Q \ll 1 \qquad \varphi_e = \varphi_1^0 \pm \frac{RT}{nF} \ln \frac{f}{a_\infty \gamma k_{\text{dif}}} \mp \frac{RT}{nF} \ln \tau_1 \qquad (\text{I. }46)$$

$$\gamma Q \gg 1 \qquad \varphi_e = \varphi_1^0 \pm \frac{RT}{nF} \ln \frac{c^0 f}{a_\infty} \qquad (\text{I. }47)$$

The upper signs in the equations (+ or −) refer to systems electrode-metal-ion of the same element in solution or systems electrode-deposit formed by the reduced form of the element, the oxidized form of the element in solution type systems. For electrodes coated with a deposit formed from the oxidized form of the element in a solution containing the corresponding reduced ions, the equilibrium potentials are described by equations (I. 45)–(I. 47) with the lower signs (− or +).

The numerical value of a_∞ is usually included in the table value of φ^0, since the latter is defined as the potential of an electrode consisting of an electroactive metal immersed in a solution with an activity of the corresponding ions equaling unity. In order to use the table values of the standard potentials in further calculations, we represent expressions (I. 45)–(I. 47) in the form:

$$\varphi_e = \varphi^0 \pm \frac{RT}{nF} \ln \frac{c^0 f}{a'_\infty [1 - \exp(-\gamma Q)]} \qquad (\text{I. }48)$$

$$\varphi_e = \varphi^0 \pm \frac{RT}{nF} \ln \frac{f}{a'_\infty \gamma k_{\text{dif}}} \mp \frac{RT}{nF} \ln \tau_1 \qquad (\text{I. }49)$$

$$\varphi_e = \varphi^0 \pm \frac{RT}{nF} \ln \frac{c^0 f}{a'_\infty} \qquad (\text{I. }50)$$

where $a'_\infty = 1$ mol/cm^3 is introduced only to maintain consistent dimensions.

Equation (I. 49) implies that when Q is sufficiently small, the equilibrium potential of the electrode is a function of the electrolytic deposition time and is independent of the concentration of the potential-determining ions in the solution. Equation (I. 50) expresses the usual dependence of the potential-determining ions in the solution, on the concentration of these ions.

The use of equations (I. 42) and (I. 48–I. 50) enables us to obtain results which fit the experimental data /3, 55/.

Chapter II

ANODIC STRIPPING VOLTAMMETRY OF METALS

The basis of anodic stripping voltammetry of metals is the electrolytic dissolution of a metal which previously had been deposited on an indifferent electrode. The deposition is usually carried out from a stirred solution or on a rotating electrode at a potential corresponding to the limiting diffusion current of the ions being discharged. If depletion of the solution is not observed, a simple relationship is obeyed /3, 56/:

$$Q = i_{\text{dif}}\tau_1$$

or on the basis of equation (I. 27)

$$Q = nFSD\frac{c^0\tau_1}{\delta} \tag{II. 1}$$

The electrolytic dissolution of a metal is carried out at a potential varying in time in a controlled fashion. In anodic stripping voltammetry, the potential is usually a linear function of the time.

The theory of anodic stripping voltammetry of metals has been developed over the last decade. Nicholson /57/, who is credited with originating anodic stripping voltammetry, treated the reversible anodic dissolution of a metallic film from the surface of an indifferent solid electrode on the assumption that the deposit forms an incomplete monolayer on a statistically homogeneous surface. An irreversible process has been described mathematically by starting from the same assumptions /58/. A mathematical description of the electrolytic dissolution of a metal from the surface of an indifferent electrode has been given for the general case by the present author /59—63/. Inasmuch as the problem is complicated and evidently insolvable in analytical form, expressions relating the parameters of a process have been obtained from analysis of the results of numerical calculations and relations describing particular cases such as the irreversible process and the reversible process on a rotating-disk electrode.

THEORY OF THE ELECTROCHEMICAL DISSOLUTION OF METALS FROM THE SURFACE OF A SOLID INDIFFERENT ELECTRODE

THE GENERAL CASE

Let us consider the electrolytic dissolution of a metal (Me) from the surface of a flat indifferent electrode into a stirred solution containing a surplus of the indifferent electrolyte under the condition that the electrode potential is a linear function of time /59—62/. The distribution of the ion concentration (c) in the layer adjacent to the electrode is described in this case by the differential equation

$$\frac{\partial c}{\partial t} = D \frac{\partial^2 c}{\partial x^2} \tag{II.2}$$

with the following boundary conditions:

$$-\frac{D}{k_S} \frac{\partial c}{\partial x}\bigg|_{x=0} = a\,\exp\left[\frac{\beta nF}{RT}(\varphi_1 - \varphi^0)\right] \cdot \exp\left(\frac{\beta nF}{RT} vt\right) -$$

$$-c(0,t)\exp\left[-\frac{\alpha nF}{RT}(\varphi_1 - \varphi^0)\right] \cdot \exp\left(-\frac{\alpha nF}{RT} vt\right) \tag{II.3}$$

$$c = c^0 \quad \text{when} \quad x \longrightarrow \infty$$

where a is the activity of the metal deposit on the electrode.

At any moment in time, a is determined by the expression:

$$a = a_\infty\left\{1 - \exp\left[-\gamma\left(Q - \int_0^t i\,dt\right)\right]\right\} \tag{II.4}$$

where

$$i = -nFSD \frac{\partial c}{\partial x}\Big|_{x=0} \qquad\qquad (\text{II. 5})$$

$\int_0^t i\,dt$ is coulometrically equivalent to the quantity of metal dissolved in time t, C, and $\left(Q - \int_0^t i\,dt \right)$ is coulometrically equivalent to the amount of metal remaining on the electrode at t, C.

At any moment, the potential is determined by the relations $\varphi = \varphi_1 + vt$.

The boundary conditions (II. 3) indicate that the transport of Me^{n+} ions to the electrode surface in the cathodic cycle and the removal of these ions into the solution bulk during dissolution take place by diffusion, and the flux of Me^{n+} ions near the electrode surface depends on the rate of the electrode reaction. In the solution bulk, the metal ions are distributed uniformly. The relationship $c(x,0) = c^0$ may serve as the initial condition /59—60/.

Equations (II. 3) and (II. 5) show that the first term in expression (II. 3) is the ionization rate and the second term is the discharge rate. Under adequate anodic polarization, the second term on the right-hand side of the equation may be neglected, and a simple equation for the dependence between the parameters of the process in which the concentration of Me^{n+} ions does not appear is obtained. Solution of this equation yields an expression for the electrochemical dissolution current for an irreversible electrode process /63/. We shall henceforth consider a process irreversible when the discharge rate of the metal ions is negligible in comparison to the ionization rate of the atoms. This is possible when the anodic polarization of the electrode is sufficient and the rate of removal of the ionization products from the electrode surface is much larger than the rate of the ionization itself. In the latter case, the rate of the discharge is small owing to the low metal ion concentration near the electrode surface.

Solution of equation (II. 2)* has yielded /61/ the dependence of the current on the electrode potential, the maximum value of the current and the potential of the polarization curve maximum, the spatial distribution of concentration for Me^{n+} ions at specified moments in time, and satisfaction of the inequalities

* See Appendix I.

characterizing the kinetics of electrochemical dissolution /59—62/:

$$-\frac{1}{k_S}\sqrt{D\frac{nF}{RT}v}\;\frac{\partial u}{\partial x}\bigg|_{x=0} \ll a'\exp\left[\frac{\beta nF}{RT}(\varphi_1-\varphi^0)+\beta\tau\right] \qquad (II.6)$$

and

$$a'\exp\left[\frac{\beta nF}{RT}(\varphi_1-\varphi^0)+\beta\tau\right] \gg u(0,\tau)\exp\left[-\frac{anF}{RT}(\varphi_1-\varphi^0)-a\tau\right] \qquad (II.7)$$

The satisfaction of inequality (II.6) indicates that the slow step is the removal of ions from the electrode surface (a reversible process), and the satisfaction of inequality (II.7) shows that the rate-determining step is electron transfer (an irreversible process).

Equations have been obtained in analytical form for specific cases of electrochemical dissolution of a metal microphase in an irreversible process as well as in a reversible process ocurring on the surface of a rotating-disk electrode.

Let us separate the parameters of a process into internal parameters peculiar to a given electrode reaction such as k_S and $\alpha(\beta)$, and external parameters such as the hydrodynamic conditions, the potential scanning rate during dissolution, and conditions for carrying out the steps preceding the anodic process. Let us now consider the effect of both types of parameters on the polarization curves for the electrochemical dissolution of a metal.

Role of the internal parameters in electrochemical dissolution

Calculated polarization curves for the stripping of a metal, which correspond to different values of the parameters (Figure 1), show that when the value of k_S decreases, the curve is shifted toward increasing τ_1, and the range of τ and, consequently, of the potentials at which dissolution occurs broadens. A positive displacement of the curves describing dissolution or an increase in the overvoltage of this process is a natural consequence of the decrease in the electrode reaction rate. When $k_S=1$ cm/sec, the rate-limiting step is the removal of ions from the electrode surface, inequality (II.6) is satisfied, and the process is reversible. When k_S equals $1 \cdot 10^{-5}$ cm/sec, the nature of the electrode-process kinetics changes. The rates of electron transfer and the removal of the reaction products into the solution bulk become commensurate, and neither inequality (II. 6) nor (II. 7) is satisfied. A further decrease in the rate constant leads to the electron transfer becoming the rate-limiting step and inequality (II. 7) is satisfied.

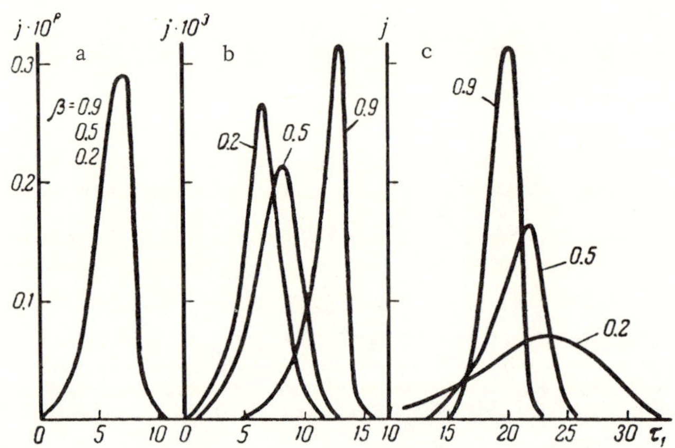

FIGURE 1. Calculated polarization curves:

a) $k_S = 1$; b) $k_S = 1 \cdot 10^{-5}$; c) $k_S = 1 \cdot 10^{-8}$ cm/sec; $c^0 = 0.25 \cdot 10^{-8}$ mol/cm³; $v = 0.1$ V/sec.

The value of the transfer coefficient, as would be expected, does not influence the shape of the curve describing a reversible process. In an irreversible process, the decrease in β is accompanied by an appreciable decrease in the maximum anodic current, clear broadening of the potential range in which dissolution occurs, and the positive displacement of the maximum current potential when the ionization of the atoms is the rate-limiting step. In the intermediate case (Figure 1, b), somewhat different behavior is observed. The maximum current potential is shifted in the positive direction when the coefficient β is increased. The maximum stripping current first decreases and then increases. This phenomenon is apparently due to a change in the nature of the electrode-process kinetics. Figure 1, b shows that the rate of the forward and reverse reactions are commensurate when β=0.2 and 0.5, and neither inequality (II. 6) nor (II. 7) is satisfied. When coefficient β is increased to 0.9, the ionization of the metal becomes the rate-limiting step, and the second inequality (II. 7) is fulfilled. The positive displacement of the curve is evidently due to such a change in the nature of the electrode process.

**Role of the external parameters in
electrochemical dissolution**

Calculated polarization curves describing the electrochemical dissolution of metal microphases /3/ (Figure 2, a) are nearly symmetric, and in contrast to the curves for macrophase dissolution (Figure 2, b), tail off more along the potential axis. The decrease after the maximum is sharper than in the stripping of macrophases. The greater tailing-off in the polarization curves for the stripping of microphases apparently is explained by the fact that the current is determined by both an increase in the electrode reaction rate as a result of the positive displacement of the potential and by the decrease in the rate caused by a decrease in the activity of the dissolving metal. In the dissolution of macrophases the activity of the metal remains constant for a fairly long time and the dissolution rate is determined only by the electrode potential until diffusion control of the current takes effect. The maximum anodic current increases with increasing concentration of the electroactive metal ions in the solution. This current is usually higher in a reversible process than in an irreversible process. The maximum current potential is constant in the dissolution of microphases and in the reversible dissolution of macrophases. An increase in the concentration of the ions of the metal forming a macrophase on the electrode or an increase in the amount of metal on the electrode which is being dissolved irreversibly is accompanied by a positive displacement of the potential of the maximum in the polarization curve.

IRREVERSIBLE ELECTROCHEMICAL
DISSOLUTION OF A METAL

In the case of an irreversible process, equation (II. 3) may be written in a short form:

$$i = nFSk_S\, a \exp\left[\frac{\beta nF}{RT}\,(\varphi - \varphi^0)\right] \qquad (II.8)$$

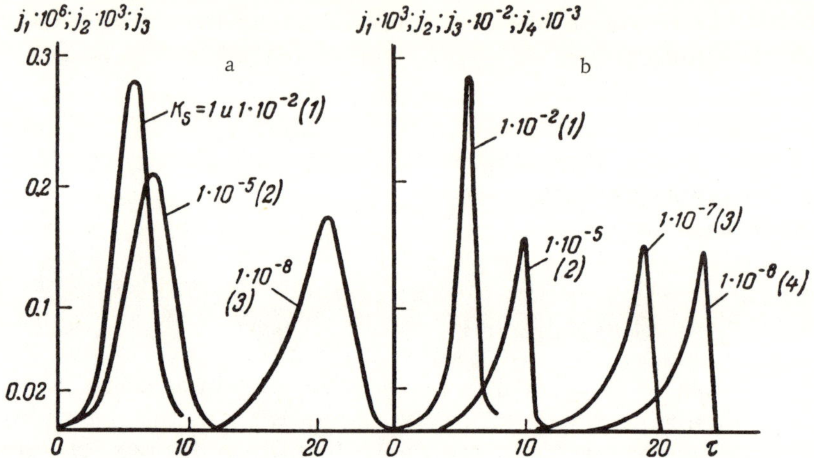

FIGURE 2. Calculated polarization curves:

a) dissolution of a microphase; $c^0 = 0.25 \cdot 10^{-8}$ mol/cm^3; $a = 0.025$; b) dissolution of a macrophase; $c^0 = 0.1 \cdot 10^{-5}$ mol/cm^3; $a = 1$; $\beta = 0.5$.

Recalling expression (II.4), we may represent equation (II.8)
in the following form:

$$i = nFSk_S a_\infty \left\{ 1 - \exp \left[-\gamma \left(Q - \int_0^t i\,d\tau \right) \right] \right\} \exp \left[\frac{\beta nF}{RT} (\varphi_1 - \varphi^0) \right] \exp \left(\frac{\beta nF}{RT} vt \right)$$

$$(II.9)$$

The solution of this equation, when the amount of deposit on
the electrode is equivalent to Q, at the starting point $t=0$, has the
form (see Appendix II):

$$i = nFSk_S a_\infty (\exp \gamma Q - 1) \times$$

$$\times \frac{\exp \left\{ \dfrac{RTSk_S a_\infty \gamma}{\beta v} \left(1 - \exp \dfrac{\beta nF}{RT} vt \right) \exp \left[\dfrac{\beta nF}{RT} (\varphi_1 - \varphi^0) \right] + \right.}{(\exp \gamma Q - 1) \exp \left\{ \dfrac{RTSk_S a_\infty \gamma}{\beta v} \left(1 - \exp \dfrac{\beta nF}{RT} vt \right) \right.}$$

$$\longrightarrow \frac{\left. \dfrac{\beta nF}{RT} (\varphi_1 - \varphi^0 + vt) \right\}}{\left. \exp \left[\dfrac{\beta nF}{RT} (\varphi_1 - \varphi^0) \right] \right\} + 1}$$

$$(II.10)$$

When $\gamma Q \ll 1$, equation (II.10) yields an expression for the
current for the electrochemical dissolution of a metal microphase:

$$t = nFSk_S \gamma Q a_\infty \exp \left\{ \frac{RTSk_S a_\infty \gamma}{\beta v} \left(1 - \exp \frac{\beta nF}{RT} vt \right) \exp \left[\frac{\beta nF}{RT} (\varphi_1 - \varphi^0) \right] + \right.$$

$$\left. + \frac{\beta nF}{RT} (\varphi_1 - \varphi^0 + vt) \right\}$$

$$(II.11)$$

The expressions for the maximum current potential, the maximum current, and the points of inflection have the form:

$$\varphi_{max} = \varphi^0 - \frac{RT}{\beta nF} \ln \frac{RT S k_S a_\infty \gamma}{\beta} + \frac{RT}{\beta nF} \ln v \qquad (II.12)$$

$$i_{max} = \frac{nF}{RT} \beta v Q \exp \left\{ \frac{RT k_S a_\infty \gamma S}{\beta v} \exp \left[\frac{\beta nF}{RT} (\varphi_1 - \varphi^0) \right] - 1 \right\} \qquad (II.13)$$

$$\varphi'_{inf} = \varphi^0 + \frac{RT}{\beta nF} \ln \zeta' - \frac{RT}{\beta nF} \ln \frac{S k_S a_\infty \gamma RT}{\beta v}$$

$$\varphi''_{inf} = \varphi^0 + \frac{RT}{\beta nF} \ln \zeta'' - \frac{RT}{\beta nF} \ln \frac{S k_S a_\infty \gamma RT}{\beta v} \qquad (II.14)$$

where φ'_{inf} and φ''_{inf} are the potentials at the points of inflection of the ascending and descending branches of the curve, $\zeta' = 0.4$, $\zeta'' = 2.61$ (see Appendix II).

The difference in the potentials at the points of inflection $(\Delta\varphi_{inf})$ equals $\Delta\varphi_{inf} = \frac{1.87}{\beta n}$.

The relationships between the currents at the points of inflection (i_n) and at the curve maximum are given by the expressions:

$$\frac{i'_n}{i_{max}} = \frac{\zeta' e}{e^{\zeta'}} = 0.72 \qquad (II.15)$$

$$\frac{i''_n}{i_{max}} = \frac{\zeta'' e}{e^{\zeta''}} = 0.53$$

Thus, the value of the maximum anodic current in the dissolution of metal microphases /3/ is directly proportional to the amount of metal on the electrode or, under fixed deposition conditions, to the concentration of the metal ions in solution (see equation (II.1)). The maximum current potential is a function of the kinetic parameters of the process (k_S and β) and the rate of varying the electrode potential. The difference between the potentials corresponding to the points of inflection of the ascending and descending branches of the curve is determined by the value of β.

REVERSIBLE ELECTROCHEMICAL DISSOLUTION OF A METAL

In a reversible process (the slow step is the removal of reaction products in the solution bulk), the electrode potential at any given moment in time is the equilibrium potential with respect to the potential-determining ion concentration near the electrode surface and to the activity of the corresponding metal on the surface:

$$\varphi = \varphi_1^0 + \frac{TR}{nF} \ln \frac{cf}{a} \tag{II.16}$$

$$\varphi^0 = \varphi_1^0 - \frac{RT}{nF} \ln a_\infty \quad \text{(see Chapter I)}$$

Let the linear scanning of the potential begin from the potential φ which is the equilibrium potential relative to the initial metal ion concentration (c^0) in the solution bulk and the metal activity on the electrode /3/, then

$$\varphi_1 = \varphi_1^0 + \frac{RT}{nF} \ln \frac{c^0 f}{a_\infty \,[1 - \exp(-\gamma Q)]} \tag{II.17}$$

For rotating-disk electrodes, the hydrodynamic conditions are such that the thickness of the diffusion layer δ is virtually constant for all points on the rotating disk and is determined by the rotation rate of the electrode, the diffusion coefficient of the reactant, and the kinematic viscosity of the medium /23/. The expression for the current may be represented as

$$i = \frac{nFSD}{\delta} (c - c^0) \tag{II.18}*$$

In the reversible dissolution of a metal, $c \gg c^0$ and c^0 may be omitted in equation (II.18).** Then

$$c = \frac{i\delta}{nFSD} \tag{II.19}$$

---

* For a nonsteady-state process such as the electrochemical dissolution of a metal under a linearly varying potential, equation (II.18) is valid only when $RT/nFv \gg \delta^2/D$, where RT/nFv is the characteristic time and δ^2/D is the relaxation time of the diffusion layer.

** For a solution of the equation without this simplifying assumption, see Appendix III.

Combining equations (II. 16), (II. 17), (II. 19) and recalling relation (II. 4), we obtain an expression for the electrode potential:

$$\varphi = \varphi_1^0 + \frac{RT}{nF}\ln\frac{c^0 f}{a_\infty\,[1-\exp(-\gamma Q)]} + vt =$$

$$= \varphi_1^0 + \frac{RT}{nF}\ln\frac{i\delta f}{nFSDa_\infty\left[1-\exp\left(-\gamma Q+\gamma\int_0^t i\,d\tau\right)\right]} \tag{II.20}$$

Solution of this equation yields the dependence of the stripping current of a metal from the surface of an indifferent electrode on the time (or the electrode potential) in the form

$$i = \frac{nFSDc^0\exp\left(\dfrac{nF}{RT}vt\right)\exp\gamma Q}{\delta\left\{\exp\gamma Q-1+\exp\left[\dfrac{SDc^0\gamma RT}{\delta[1-\exp(-\gamma Q)]v}\left(\exp\dfrac{nF}{RT}vt-1\right)\right]\right\}} \tag{II.21}$$

When $\gamma Q \ll 1$, solution of equation (II. 20) reduces to:

$$i = \frac{Q}{\tau_1}\exp\left[\frac{RT}{nFv\tau_1}\left(1-\exp\frac{nF}{RT}vt\right)\right]\exp\frac{nF}{RT}vt \tag{II.22}$$

We may now easily obtain a relationship for the potential of the polarization curve maximum and the maximum current value for the electrochemical dissolution of a metal in the form

$$\varphi_{\max} = \varphi_e + vt_{\max} = \varphi_e + \frac{RT}{nF}\ln\frac{nF}{RT}\tau_1 + \frac{RT}{nF}\ln v \tag{II.23}$$

$$i_{\max} = \frac{nF}{RT}vQ\exp\left(\frac{RT}{nFv\tau_1}-1\right) \tag{II.24}$$

Thus, the reversible stripping of a metal from the surface of a rotating-disk electrode is also described by a polarization curve with a maximum current which is directly proportional to the amount of metal on the electrode or according to formula (II. 1) to the metal ion concentration in solution. The maximum current and the potential at the curve maximum depend on the electrode potential scanning rate.

Comparison of the results of numerical calculations with calculations based on equations (II. 12), (II. 13), (II. 23), and (II. 24) confirms the following relationships for the parameters of metal stripping:

1. Reversible process
A. Microphase:
rotating electrode

$$i_{max} = \frac{nF}{RT}\, vQ \exp\left(\frac{RT}{nFv\tau_1} - 1\right)^{*}$$

$$\varphi_{max} = \varphi_p + \frac{RT}{nF} \ln \frac{nF}{RT}\,\tau_1 + \frac{RT}{nF} \ln v$$

stationary electrode

$$i_{max} = 0.79\, \frac{nF}{RT}\, vQ \exp\left(\frac{\delta}{\sqrt{\dfrac{nF}{RT} Dv\ \tau_1}} - 1\right)^{*}$$

$$\varphi_{max} = \varphi_e + \frac{RT}{nF} \ln \frac{\tau_1 \sqrt{\dfrac{nF}{RT} D}}{\delta} + \frac{RT}{2nF} \ln v$$

B. Macrophase:
stationary electrode

$$i_{max} = -0.93\, \frac{nF}{RT}\, vQ' + 0.93\, \frac{nF}{RT}\, vQ^{**}$$

2. Irreversible process
A. Microphase:

$$i_{max} = \frac{nF}{RT}\, \beta vQ \exp\left\{ \frac{RTSk_S\, a_\infty\, \gamma}{\beta v} \exp\left[\frac{\beta nF}{RT}(\varphi_1 - \varphi^0) \right] - 1 \right\}^{*}$$

$$\varphi_{max} = \varphi^0 + \frac{RT}{\beta nF} \ln \frac{\beta}{RTSk_S\, a_\infty\, \gamma} + \frac{RT}{\beta nF} \ln v$$

B. Macrophase:

$$i_{max} = -0.93\, \frac{nF}{RT}\, \beta vQ' + 0.835\, \frac{nF}{RT}\, \beta vQ^{**}$$

The equations obtained imply that the polarization curves describing the electrochemical dissolution of a metal have characteristic common features, namely, the existence of a maximum

---

* $RT/nFv\tau_1$; $\delta / \sqrt{\dfrac{nF}{RT} Dv\ \tau_1}$; $\dfrac{RTSk_S\, a_\infty\, \gamma}{\beta v} \exp\left[\dfrac{\beta nF}{RT}(\varphi_1 - \varphi^6) \right] \ll 1$ the corresponding values of the exponents are about e^{-1}.

** Q' is the point of intersection between the extension of the linear section corresponding to large Q and the abscissa.

current directly proportional to the amount of metal on the electrode and the metal ion concentration in the solution upon dissolution of a microphase and a linear relationship between these values in the dissolution of a macrophase. This behavior allows the use of electrodeposition with the subsequent recording of the dissolution current as an analytic method.

The position of the curve relative to the potential axis, the nature of the dependence of the maximum current potential in the curve on the potential scanning rate, the slope of the linear sections of the curves in the i_{max}—Q graphs, and the potential difference at the points of inflection on the ascending and descending branches of the curve enable us to determine the kinetic parameters of the process.

USE OF ANODIC STRIPPING VOLTAMMETRY OF METALS FOR STUDYING THE KINETICS OF ELECTRODE PROCESSES

The variation in the polarization curves for the electrochemical dissolution of metals as a function of the kinetics of the electrode process makes it possible to use anodic stripping voltammetry of metals to study the kinetics of the process. Calculated values of several parameters for a stationary electrode (at 25°C) are:

	Reversible process	Irreversible process
di_{max}/dQ		
Microphase	11.2 nv*	14.2 βnv
Macrophase	35.8 nv	32 βnv
$(di_{max}/dQ)_{macr} : (di_{max}/dQ)_{micr}$	3.2	2.3
$(di_{max}/dQ)_{rot} : (di_{max}/dQ)_{sta}$	1.27	1
$d\varphi_{max}/d\log v$	0.030/n**	0.059/βn
$\Delta(nFvt_{inf}/RT)$		
Microphase	—	1.87/β
Macrophase	—	0.71/β

 * For a rotating electrode this value equals 14.2 nv.

** For a rotating electrode this value equals 0.059 /n.

Criteria for reversibility of the process are:

1) the derivative of the maximum anodic current with respect to q and the ratio of these derivatives for the macro- and microphases;

2) the ratio of the maximum current of electrolytic dissolution of an identical amount of metal from the surface of rotating and stationary electrodes;

3) the derivative of the maximum current potential with respect to the logarithm of the potential scanning rate.

In a reversible process, the ratio of the derivatives of the maximum current from the stripping of a metal microphase from the surface of rotating and stationary electrodes with respect to the amount of deposit on the electrode is greater than unity. In an irreversible process, this ratio equals unity since the discharge reaction rate is small and the value of the metal ion concentration in the region adjacent to the electrode does not appear in the equation for the current. Therefore, stirring the solution or rotating the electrode does not affect the polarization curves. The maximum current in the irreversible process is lower than in the reversible process when the maximum current potential is displaced only slightly with change in the scanning rate. This displacement depends on the hydrodynamic conditions.

Another criterion for reversibility of the process is the nature of the time-dependence of the electrochemical dissolution current of a metal macrophase at constant electrode potential. The current flowing through the electrode-solution interface at constant potential equals

$$i(\varphi) = i_+(\varphi) - i_-(\varphi) \tag{II.25}$$

The value of i_- depends on the metal ion concentration in the region adjacent to the electrode. This concentration is diffusion-controlled. Thus, for stationary electrodes, i_- is a function of time /5/. The ionization rate of metal atoms in a macrophase is not time-dependent in a certain range since the atoms participating in this step are on the electrode surface and the activity of the metal does not begin to change immediately. Thus, the time-dependence of the current from the electrochemical dissolution of a metal macrophase at constant potential when $i_+ \gg i_-$ (irreversible process) should vanish over a certain interval. On the other hand, when $i_- \gg i_+$, the time-dependence should be close to a hyperbolic. When t is sufficiently large, the current in both cases vanishes due to the complete removal of the metal from the electrode surface.

KINETICS OF THE ELECTROCHEMICAL DISSOLUTION OF SILVER

The electrochemical dissolution of silver from the surface of graphite /62/ and platinum /57, 62/ electrodes has been studied. These experiments were carried out in two steps, namely, deposition of silver on a rotating electrode at a potential corresponding to the limiting diffusion current of the metal ions (the deposition time was 60 sec and the potential 0.6 V relative to a mercurous sulfate electrode) and the electrochemical dissolution with a linearly varying potential of the rotating or stationary electrode. After the first step, the electrode was set close to the equilibrium potential. Graphite (type I)* and platinum disk-electrodes, of 2 mm diameter, were used as the working electrodes.

The platinum electrode sealed in a glass tube was ground, polished, and fired. The experiments were performed at 20°C. The solutions were prepared in doubly distilled water and surface-active impurities were removed in a column filled with charcoal. The charcoal was first washed for two weeks in a soxhlet apparatus with boiling hydrochloric acid, for three days with water and then calcined at red heat temperature in a nitrogen flow in a quartz vessel. Fresh silver nitrate solutions were prepared daily. Oxygen was removed by a nitrogen flow washed by a vanadium (II) sulfate solution which was continuously regenerated by a zinc amalgam placed at the bottom of the washing vessels. The gas was further washed by a sodium plumbite solution and water. The amount of metal deposited in the precipitation was determined from the area between the stripping polarization curve and the residual current line.

An 02TsLA oscillographic polarograph with three additional condensers of $4\,\mu F$ capacity each in the scanning block was used as a source and recorder of the current. This modification allowed the following additional scanning rates: 0.01, 0.02, 0.04, 0.085, 0.17, 0.66, and 1.33 V/sec. The linearity of the varying potential of the working electrode was monitored using a mercurous sulfate reference electrode and an N-373 millivoltmeter with an internal resistance of $5\,M\Omega$.

Polarization curves for the stripping of a silver microphase (a) from the surface of rotating and subdued graphite and platinum electrodes and of a macrophase (b) from the surface of a subdued graphite electrode are presented in Figure 3. The maximum current for the electrochemical dissolution of silver from the surface of a rotating electrode is shown to be somewhat higher and a negative

---

* Methods for preparing graphite electrodes are described in Chapter V.

shift in the potential of the maximum is noted. As expected, the curves obtained for graphite and platinum electrodes are similar.

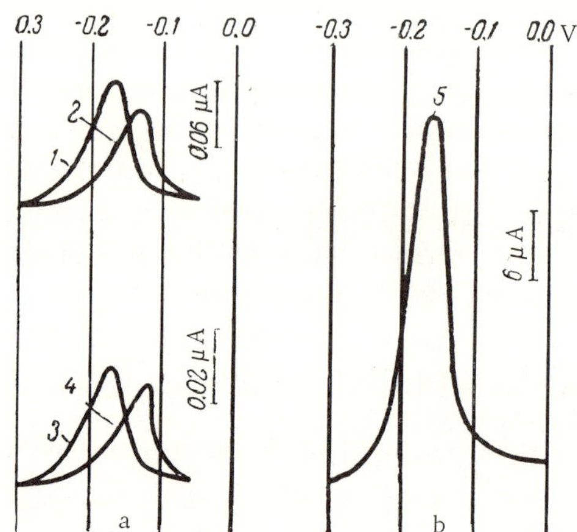

FIGURE 3. Polarization curves for the stripping ($v = 0.01$ V/sec) of a silver microphase (a) and macrophase (b):

1, 2) from a rotating and a subdued graphite electrode; 3, 4) from a rotating and a subdued platinum electrode (the deposition was from a solution 1 N in KNO_3 and $5 \cdot 10^{-7}$ N in $AgNO_3$); 5) from a subdued graphite electrode (the deposition was from a solution 1 N in KNO_3 and $5 \cdot 10^{-5}$ N in $AgNO_3$).

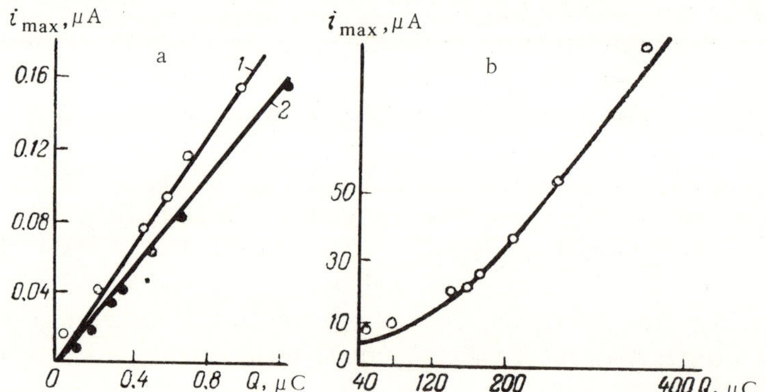

FIGURE 4. Dependence of the maximum stripping current of silver on Q:

a) dissolution of the microphase from a rotating (1) and a subdued (2) electrode; b) dissolution of the macrophase from a subdued electrode.

The maximum dissolution current is a function of the amount of material on the electrode. The nature of the dependence (which is similar to the dependence of the maximum on the silver ion concentration in the solution) is illustrated in Figure 4. The plot of the dependence of the maximum stripping current of the silver micro-phase from the surface of a rotating-disk graphite electrode on the amount of silver is located above the corresponding plot for the sub-dued electrode.

The maximum stripping current of a silver macrophase (Fig-ure 4b) is linearly dependent on the amount of metal on the elec-trode and the slope of the linear section of this part of the plot is greater than that of line 2 (Figure 4a).

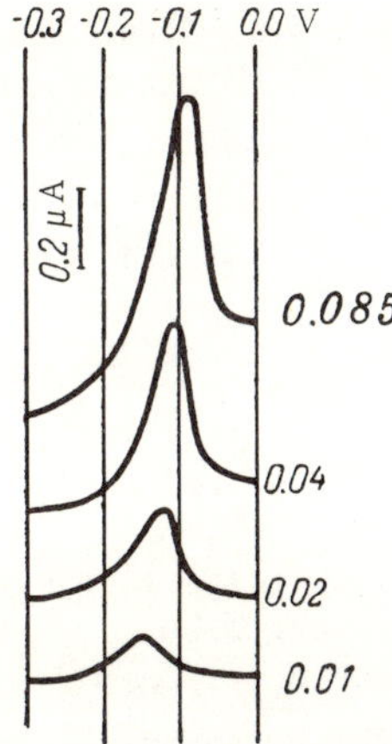

FIGURE 5. Polarization curves for the stripping of silver obtained at different potential scanning rates (numbers on the plots in V/sec).

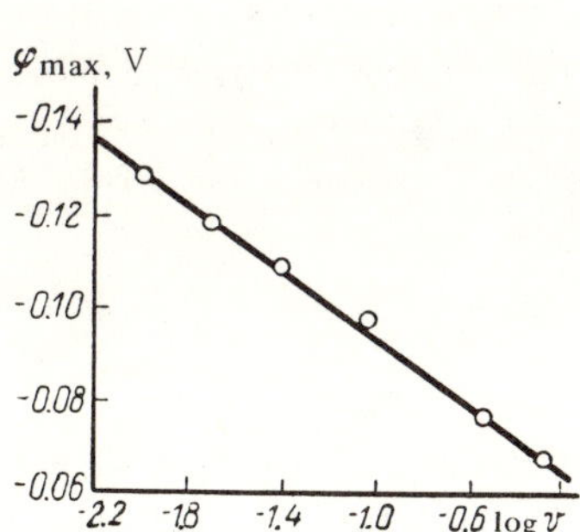

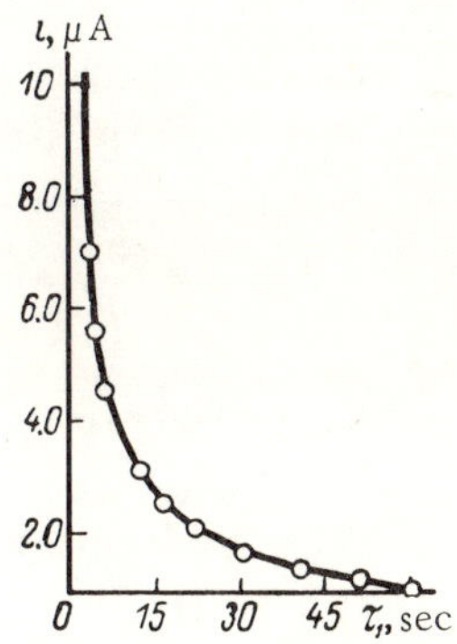

FIGURE 6. Dependence of the maximum stripping current potential on the logarithm of the electrode potential scanning rate.

FIGURE 7. Time-dependence of the silver stripping current at 0.1 V.

Polarization curves for silver stripping from the surface of a subdued graphite electrode for different potential scanning rates are given in Figure 5. The maximum stripping current increases with increasing potential scanning rate. The dependence of the maximum current potential on the logarithm of the scanning rate is depicted in Figure 6. The maximum current potential undergoes a positive displacement when v is increased. The time-dependence of the stripping current of the silver macrophase at constant potential approximates a hyperbole (Figure 7).

Here is a comparison of the values which characterize the kinetics of silver stripping at $v = 0.01$ V/sec on a stationary electrode calculated for a reversible electrochemical reaction and those obtained from experimental data:

	Calculated	Experimental
di_{max}/dQ		
microphase	0.11*	0.12*
macrophase	0.35	0.33
$(di_{max}/dQ)_{macr} : (di_{max}/dQ)_{micr}$	3.2	2.8
$(di_{max}/dQ)_{rot} : (di_{max}/dQ)_{sta}$	1.27	1.23
$d\varphi_{max}/d \log v$	0.030	0.037
$i = f(t)$ when $\varphi = const$	hyperbolic	hyperbolic

* For rotating electrodes, the calculated and experimental values equal 0.142 and 0.146, respectively.

The data given show that the experimental values are close to those calculated for a reversible electrode process. This result indicates that the reaction $Ag \rightleftharpoons Ag^+ + e$ proceeds rapidly. The rate-limiting step is the removal of Ag^+ ions from the electrode surface.

KINETICS OF NICKEL STRIPPING

The electrochemical dissolution of nickel from the surface of a graphite (type I) electrode has been studied in our laboratory /63/. The experimental technique is similar to that described in the preceding section. The polarization curves for the stripping of nickel from the surface of a subdued (1) and a rotating (2) electrode recorded for a linearly varying potential and the time-dependence of

the nickel macrophase stripping current are given in Figures 8 and 9.
The figures show that rotation of the electrode does not affect the
polarization curve and that the stripping current at constant potential
is virtually time-independent in a certain range. This result indicates that the process is irreversible and in agreement with other
works /64, 65/.

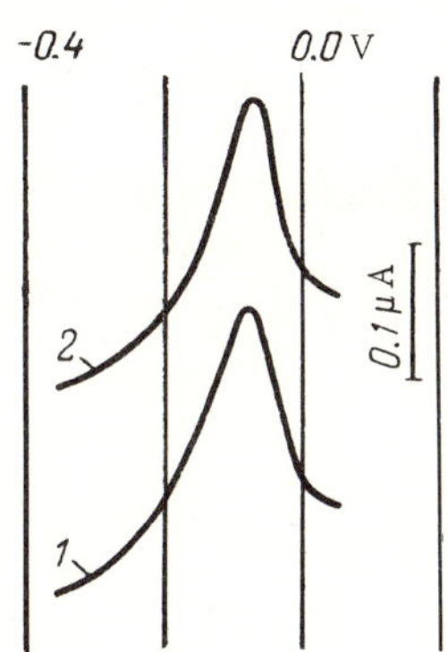

FIGURE 8. Polarization curves for the stripping
($v = 0.02$ V/sec) of nickel at $\tau_1 = 1$ min,
$\varphi_{el} = -1.2$ V from a solution 1 M in KNO_3,
0.001 M in HNO_3, and $3 \cdot 10^{-5}$ M in $Ni(NO_3)_2$
from a subdued (1) and a rotating (2) electrode.

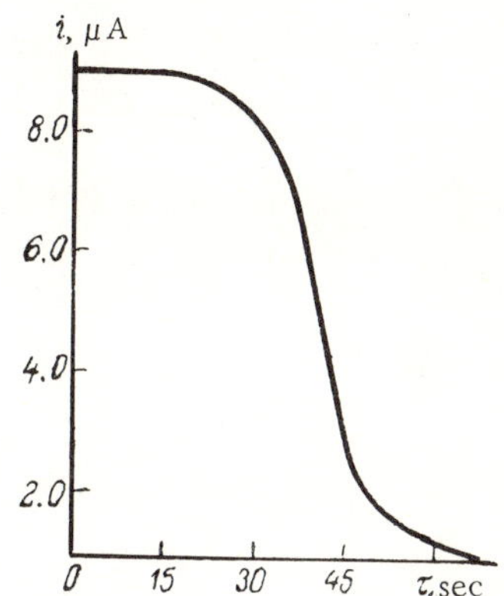

FIGURE 9. Time-dependence of the
nickel stripping current at -0.1 V.

The dependence of the maximum stripping current of nickel
deposited from a solution of potassium nitrate acidified with nitrous
acid on the amount of metal on the electrode, the nickel ion concentration in the solution, and the deposition time is shown in Figure 10.
The maximum stripping current for the metal microphase is directly
proportional to the amount of metal on the electrode, the concentration of metal ions in the solution, and the deposition time. There is
also a linear dependence between the maximum current for stripping
of the nickel macrophase and the amount of metal on the electrode.
The slopes of the linear sections of the plot equal 0.13 and 0.30. The
ratio of these values is 2.3, which agrees with the value calculated
for the irreversible process.

The effect of the potential scanning rate on the polarization
curves for the dissolution of nickel in a potassium nitrate solution
is illustrated in Figures 11 and 12. The figures show that an increase in the potential scanning rate is accompanied by a positive
displacement of the potential of the curve maximum. This displacement is far more pronounced than in the reversible dissolution of

silver /62/.   The function $\varphi_{max} = f(\log v)$ is a straight line with slope equal to 0.13.

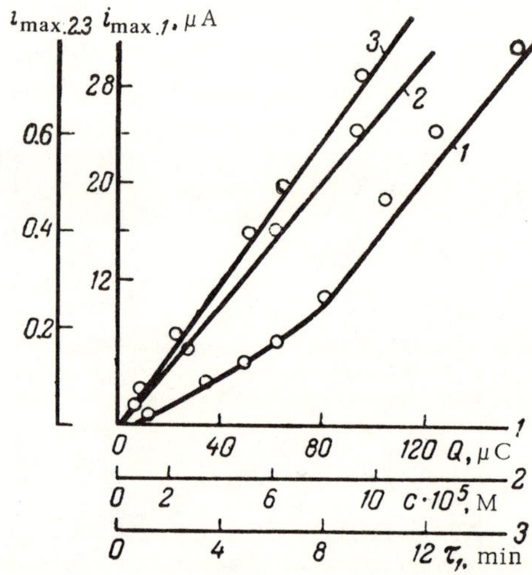

FIGURE 10. Dependence of the maximum dissolution current of nickel on the amount of nickel on the electrode (1), the concentration of nickel ions in the solution (2), and the deposition time (3).

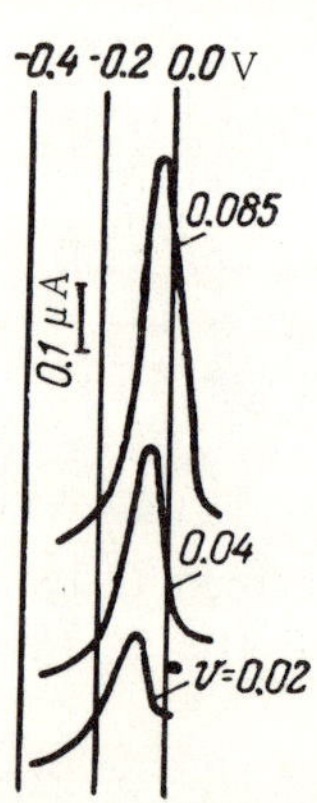

FIGURE 11. Polarization curves for the stripping of nickel deposited at $\tau_1 = 1\,min$ and $\varphi_{el} = -1.2\,V$ from a solution 1 M in KNO_3, 0.001 M in HNO_3, and $3 \cdot 10^{-5}$ in $Ni(NO_3)_2$.

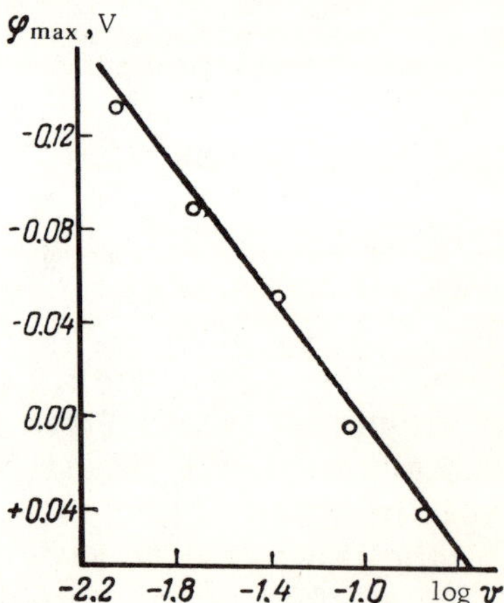

FIGURE 12. Dependence of the potential of the polarization curve maximum for the dissolution of nickel on the logarithm of the potential scanning rate (for conditions see Figure 11).

The values of β_n calculated from the dependence of the maximum dissolution current on the amount of nickel on the electrode and the dependence of the maximum current potential on the logarithm of the potential scanning rate were equal to 0.46 in both cases.

Transforming equation (II. 12) and using the data in Figure 12, we can calculate an approximate value for the rate constant of the electrode process, k_S, from the formula

$$k_S = \frac{Bv}{nFa_\infty \gamma S} \exp\left[-B\left(\varphi_{max} - \varphi^0\right)\right] \tag{II. 26}$$

where

$$B = \frac{\beta nF}{RT}; \qquad a_\infty = \frac{d}{M}; \qquad \gamma = \frac{1}{Q'}$$

where Q' is the coulombic equivalent to the amount of metal in the transition from micro- to macrophase (determined by extrapolation of the linear section of curve 1 (see Figure 10) to the abscissa) /3/, and φ_{max} is the empirical value for the potential at the polarization curve maximum.

The value of k_S for the reaction $Ni \rightleftharpoons Ni^{2+} + 2e$ is found equal to $4 \cdot 10^{-10}$ cm/sec. Sathyanarayna /64/ obtained $2.4 \cdot 10^{-9}$ and $4 \cdot 10^{-9}$ cm/sec for a similar reaction on a mercury capillary electrode in an 0.25 M solution of $NaClO_4$. Delahay and Mattax /65/ found that $k_S = 2.6 \cdot 10^{-10}$ cm/sec, and Morinaga /66/ reported a value of $k_S = 10^{-10}$ cm/sec. In view of the different conditions for the mercury /64—66/ and graphite /63/ electrodes and the completely different theoretical approach, we should conclude that the value of k_S obtained from the anodic stripping voltammetry measurements agrees well with the corresponding values obtained by the other methods.

APPLICATION OF ANODIC STRIPPING
VOLTAMMETRY OF METALS IN ANALYSIS

The dependence existing between the maximum current of the stripping of a metal deposited on an indifferent electrode and the concentration of ions of this metal in the solution enables us to use anodic stripping voltammetry in analysis. The ability to determine elements by anodic stripping voltammetry of metals is limited by the working potential range of the indifferent electrode being employed. From this point of view, specially prepared graphite electrodes are most suitable, as they are electrochemically stable and the discharge-ionization reactions of hydrogen and oxygen occur on these electrodes with a high overvoltage. Thus, in neutral medium, there is a virtually free range of potentials from +0.9 to -1.2 V relative to the saturated calomel electrode which in acid medium is positively shifted and in alkaline medium, negatively shifted. Therefore it is possible to determine both noble metals and metals shifted in a voltage range toward negative potentials. Methods have been developed for determining gold, silver, mercury, copper, bismuth, antimony, lead, tin, nickel, cobalt, thallium, indium, cadmium, and iron.

Analysis is not difficult for ions of electropositive metals in the presence of electronegative elements. The simultaneous determination of several metals is possible when the potentials of the maxima in the respective electrochemical oxidation polarization curves differ by no less than 0.1—0.2 V and the metals do not form solid solutions or intermetallic compounds. However, these limitations can be removed in most cases by selecting a special background, using complexation and introducing a small amount of a divalent mercury salt into the solution being analyzed.

To determine the concentration of ions of most metals, except for the most electronegative metals, we may use pyrolytic vitreous carbon electrodes impregnated with an epoxy resin and polyethylenepolyamine (type I) or with wax and polyethylene (type III), paste electrodes (type II), pyrolytic electrodes or vitreous carbon electrodes. For work in the negative potential range, we should choose more hydrophilic electrodes, such as those impregnated with an epoxy resin and polyethylenepolyamine. In the potential range from +0.3 to +1.2 V, especially in acid medium, the use of electrodes impregnated with a mixture of polyethylene and paraffin wax is preferable. Recent investigations indicate that the most universal electrodes are from pyro- and vitreous carbon.

The ions of the metals to be described are reduced to the elementary state on graphite electrodes. In all cases, unless stipulated otherwise, the cathodic polarization curves are well defined and have the form of conventional polarograms. The limiting cathodic current is directly proportional to the ion concentration.

The electrolysis of solutions at an electrode potential adequate to reduce the metal ions is accompanied by the formation of a metal deposit on the electrode. The quantity of deposit and the maximum current for the electrochemical oxidation of the deposit increase as the electrode potential in the concentrating step is shifted in the negative direction usually up to a value 0.2—0.3 V more negative than the half-wave potential of the cathodic polarogram. At more negative potential values, the anodic current is independent of the electrode potential in the preliminary step.   The maximum stripping current of a metal concentrated under constant hydrodynamic conditions and at the limiting diffusion current is proportional to the concentration of the metal ions in the solution and to the electrodeposition time.

We shall treat the most complicated cases of analysis which require special sample preparation.   The method of analysis to determine the concentration of the ions of electropositive metals in materials containing electronegative elements may easily be chosen on the basis of the general characteristics of the metal and is hardly treated here.   To analyze certain metals, we may use amalgam polarography with accumulation by replacing the conventional hanging mercury drop electrode with a graphite electrode and adding $Hg(NO_3)_2$ to a concentration of $10^{-6}-10^{-5}$ M.

DETERMINATION OF ELEMENTS

Gold

The half-wave potential of the cathodic polarogram of gold on a background of dilute nitric acid is close to +0.75 V.*   The electrochemical dissolution of the metal occurs in the +0.75—+0.95 V potential range.   Discharge of the supporting electrolyte ions is not observed.

The maximum anodic current increases as the electrode potential in the concentrating step goes from +0.7 to -0.2 V.   A further increase in the cathodic polarization does not alter the anodic polarograms.

Reduction of the metal ions and oxidation of the metal occur in the more negative potential range when hydrochloric acid serves as the supporting electrolyte, since fairly stable gold complexes form in hydrochloric acid.   Concentrating the metal from a hydrochloric acid solution onto the electrode begins at a potential of +0.2 V.   However, the process proceeds slowly at this potential.   The maximum current of the electrochemical dissolution of metal concentrated at potentials more negative than -0.2 V is independent of the potential. The cathodic polarogram half-wave potentials and the potentials of the anodic curve maxima are very close.   The sensitivity of gold determination /185/ in hydrochloric acid is 10^{-10} wt. %.

Techniques have been proposed for the determination of gold in antimony, electroplatings, and plant ash /67/ with the use of graphite (type II) and saturated calomel electrodes.

---

* Again the potentials are given relative to the saturated calomel electrode.

Determination of gold in antimony /67/. First, an 0.5—1 g powdered antimony sample is dissolved in 15 ml of a nitric and hydrochloric acid mixture (1 : 3). After the dissolution of the sample 5 ml hydrobromic acid is added, and the solution is evaporated to dryness over a water bath. Dissolving the dry residue in 8—10 ml of hydrobromic acid and evaporation is repeated three or four times in order to completely remove the antimony. The residue is dissolved in 10 ml 1 M hydrochloric acid and the solution is transferred to the electrolysis cell. Oxygen is removed using an inert gas and the graphite electrode is polarized at + 1.3 V for 15 min. Electrolysis of the stirred solution is carried out for 30 min at + 0.2 V. The stirring is stopped and the solution allowed to stand for 15—20 sec, after which the anodic polarization curve is recorded for potentials ranging from + 0.2 to 1.3 V.

The anodic polarization curve of the background is recorded beforehand under the same conditions. The concentration of the gold is found from a known addition.

In the determination of $1 \cdot 10^{-4}\%$ gold, the coefficient of variation is 5—7%.

Determination of gold in platings on tungsten and molybdenum /67/. First, an 0.15—0.20 mg sample is dissolved in 5 ml of a nitric and hydrochloric acid mixture (1:3). The solution is evaporated over a water bath to the moist salt state and then two or three crystals of sodium chloride are added in the determination of gold in a plating on tungsten or 1 ml 1 M hydrochloric acid is added in the determination of gold in a plating on molybdenum. The solution is transferred into a 25 ml volumetric flask and brought up to the line with 1 M hydrochloric acid. The analysis is continued as outlined in the preceding method.

Determination of gold in plant ash /67/. First, 0.05—0.1 g of well-ground plant ash calcined at 400—500°C is dissolved in 5 ml of a mixture of nitric and hydrochloric acid (1:3) in a 25 ml glass crucible. The solution is evaporated over a water bath to the moist salt state and two or three crystals of sodium chloride are added. The moist residue is dissolved in 1 M hydrochloric acid, transferred to a 100 ml volumetric flask, and brought up to the line with 1 M HCl. The analysis is then continued as outlined in the determination of gold in antimony.

Silver

The cathodic and anodic polarization curves of silver in a supporting electrolyte solution containing nitric acid and 0.1 M potassium nitrate (pH = 2) are presented in Figures 13 and 14. The cathodic current of silver reaches a limiting value at -0.4 V relative to the saturated mercurous sulfate electrode. The maximum anodic current increases as the electrode moves in the cathodic cycle from -0.3 to -0.6 V relative to the mercurous sulfate electrode.

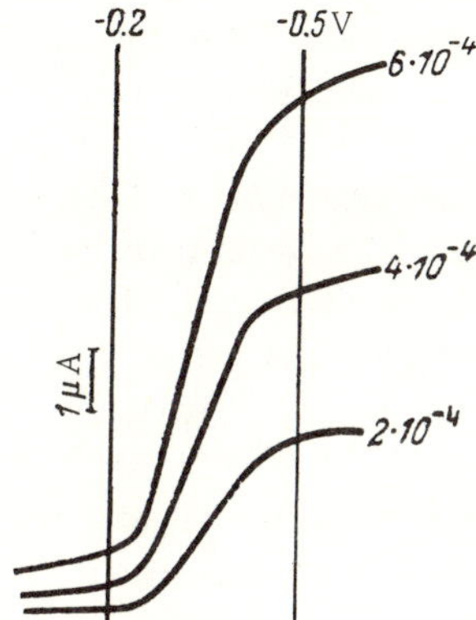

FIGURE 13. Cathodic polarization curves of silver at different Ag^+ concentrations (numbers on the curves: g · ion/liter).

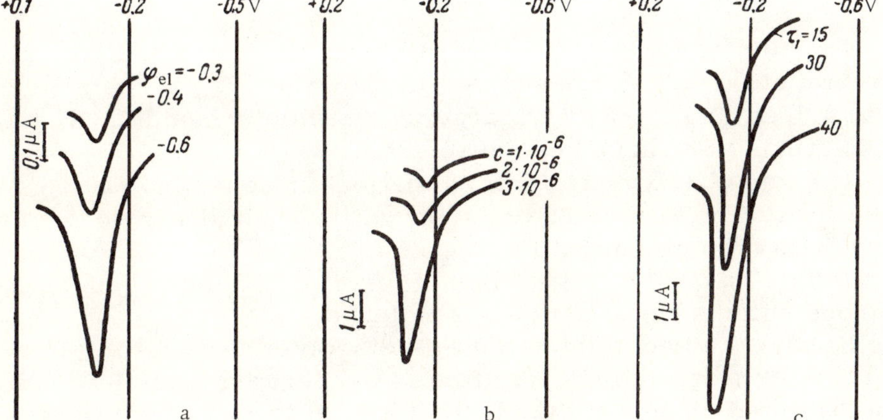

FIGURE 14. Polarization curves ($S = 0.01\,cm^2$) for the stripping of silver at $v = 0.018$ V/sec (a) and 0.125 V/sec (b, c) deposited under different conditions:

a) $(Ag^+) = 1 \cdot 10^{-6}$ g · ion/liter, $\tau_1 = 1$ min, the values of φ_{el} are the numbers on the curves (V);
b) $\tau_1 = 3$ min, $\varphi_{el} = -0.6$ V, the concentrations of Ag^+ are the numbers on the curves (g · ion/liter);
c) $(Ag^+) = 5 \cdot 10^{-7}$ g · ion/liter, $\varphi_{el} = -0.6$ V, the values of τ_1 are the numbers on the curves (min).

Figure 14 shows that the maximum anodic current increases with increasing silver ion concentration, and deposition time is faster than would have been expected for a directly proportional dependence. This phenomenon is evidently a result of the transition from the micro- to the macrophase of the metal on the electrode surface under these conditions which is accompanied by a significant increase in the maximum stripping current.

Well defined cathodic and anodic polarization curves are encountered for electrode processes involving complex silver ions (chloride and thiocyanate). The anodic curves have one maximum in a broad concentration range. The cathodic waves are strongly shifted toward negative potentials. Thus, the silver ions from 4 M hydrochloric acid solutions are reduced at potentials more negative than -0.4 V relative to the saturated calomel electrode and the half-wave potential for the reduction of the silver thiocyanate complex from a 1 M potassium thiocyanate solution is about -0.6 V relative to the saturated calomel electrode.

When a sulfuric acid background is used, the limiting cathodic current is attained at -0.1 V and electrochemical dissolution occurs at from +0.2 to +0.4 V relative to the saturated calomel electrode.

Differential anodic polarization curves have been obtained for silver deposited from a solution 0.01 M in ammonium hydroxide, 0.001 M in ammonium nitrate, and 10^{-10}—10^{-7} M in silver nitrate /68/.

Techniques for determining silver in salts of alkali and alkaline earth metals /69—71/, lead /72/, in thiourea /72/, in metallic cadmium, lead /73, 74—79/, tin, arsenic, gallium arsenide /68/, and electroplatings /68/ have been described.

The sensitivity of the determinations reaches $2.5 \cdot 10^{-9}$ g·ion/lite The ions of metals more electronegative than silver do not interfere with the determination.

Determination of silver in cadmium. * Graphite (type I) and mercurous sulfate electrodes are employed.

First, a 2 g cadmium sample is dissolved in 15 ml nitric acid (1:1) with heating. The solution is evaporated, diluted with water to 15 ml and placed in the electrolytic cell from which oxygen has been removed by an inert gas.

The connecting vessel and the electrolytic bridge are filled with 1 M nitric acid. Electrolysis of the stirred solution is carried out for 15 min at -0.5 V. The stirring is halted and the anodic polarization curve is recorded. The concentration of silver is found from a known addition.

In determining $2 \cdot 10^{-5}$% silver in cadmium, the coefficient of variation is 18—20%.

---

* The methods for determining silver in cadmium and tin were proposed by N.K.Kiva and Kh.Z.Brainina.

Determination of silver in tin.   Graphite (type I) and mercurous sulfate electrodes are used.

First, 1 g tin shavings are dissolved in 5 ml concentrated sulfuric acid.   The volume is brought up to 20 ml with distilled water and the determination is continued as in the preceding method.

In determining $1 \cdot 10^{-4}\%$ silver in tin, the coefficient of variation is 12—15%.

Determination of silver in lead /74/.   M e t h o d I.   Vitreous carbon and saturated calomel electrodes are employed.

First, 0.1—1 g lead is dissolved in nitric acid, the solution is twice evaporated to dryness, and the residue is dissolved in 20 ml 0.5 M potassium nitrate.   Electrolysis of the stirred solution is carried out for 30 min at -0.1 V.   The stirring is halted and the anodic po-larization curve is recorded at from 0 to +0.8 V.   The silver con-centration is found from a known addition.

M e t h o d II /79/.   Graphite (type I) and mercurous sulfate electrodes are employed.

First, an 0.1—0.2 g lead sample is dissolved in 10 ml nitric acid (1:1).   The solution is evaporated to dryness and the residue is dissolved in 15—20 ml 0.05 M potassium thiocyanate with gentle heating.   The solution is transferred to a heat-resistant electrolytic cell at 40°C and argon is passed through for 7—10 min to remove oxygen.   The electrolysis of the stirred solution is carried out for 5—10 min at -0.8 V.   The stirring is stopped and the anodic polari-zation curve is recorded.   The maximum dissolution current of sil-ver is at a potential of about -0.4 V.   The silver concentration is found from a known addition.

The sensitivity of the determination found from the 3σ test is $1 \cdot 10^{-9}$ g·ion/liter ($2 \cdot 10^{-6}\%$ for an 0.1 g lead sample and a 15 min deposition time).   The coefficient of variation is 22%.   The mean square error of the determination of $2 \cdot 10^{-5}\%$ silver in lead is 7%.

Determination of silver in arsenic and gallium arsenide /68/. Graphite (type II) and saturated calomel electrodes* are employed.

First, an 0.10—0.15 g arsenic or gallium arsenide sample is dissolved in a mixture of 2 ml concentrated nitric acid and 4 ml concentrated hydrobromic acid in a 25 ml quartz dish.   The solution is evaporated over a water bath almost to dryness, 5 ml concentrated hydrobromic acid is added, and the solution is again evaporated to dryness.   This operation is repeated three or four times to remove the arsenic.   Separation of gallium is not required, since it does not interfere with the determination of silver.   The residue is treated

---

* An electrolytic cell should be employed in which the igniting electrode is separated by an intermediate vessel. In order to avoid precipitation of silver by chloride ions, it is preferable to use a mercurous sulfate electrode as the counter electrode.

with the supporting electrolyte solution (0.01 M in NH_4OH and
0.001 M in NH_4NO_3) and transferred to a 10 ml volumetric flask.
An 0.1 M ammonium hydroxide solution is added dropwise to bring
the solution to pH = 9, and the volume is brought up to 10 ml with the
supporting electrolyte solution.   Then, the solution is transferred to
the electrolytic cell, and oxygen is removed by bubbling nitrogen.

The electrode is subjected to anodic polarization for five minute
at +0.5 V.   Then, electrolysis of the stirred solution is carried out
for ten minutes at -0.4 V.   The stirring is terminated and the differential anodic curve recorded.   The concentration of silver is
found from a known addition.

In a determination of $2 \cdot 10^{-7}\%$ silver, the coefficient of variation
is 15%.

Determination of silver in electroplatings /68/.   To determine
silver in platings on molybdenum, an 0.5 mg sample of silver-plated
wire weighed to an accuracy of six significant figures is dissolved
in 5 ml concentrated nitric acid in a quartz dish.   The solution is
evaporated almost to dryness and a solution 0.01 M in NH_4OH and
0.0001 M in NH_4NO_3 is added as the supporting electrolyte.   The
solution is transferred to a 100 ml volumetric flask, brought to pH =
and filled up to the calibration mark with the supporting electrolyte
solution.   The analysis is continued as outlined in the preceding
method.

To determine silver on a molybdenum wire attached to a copper
part, the whole part is weighed analytically and dissolved in 10 ml
concentrated nitric acid in a quartz dish.   The solution is evaporate
to dryness over a water bath, the supporting electrolyte solution
added, the solution transferred to a 100 ml volumetric flask, pH = 9
established, and the solution brought up to the mark with the suppor
ing electrolyte solution.   Then 1 ml of the solution is placed in a
100 ml volumetric flask and again brought up to the mark with the
supporting electrolyte solution.   Finally, 1 ml of the solution is
transferred to the electrolytic cell, 9 ml of the supporting electrolyt
solution added, and the solution is polarized under the conditions ou
lined above.

The error in the determination of 2—3% Ag is 15%.

The determination of silver is not disturbed by a hundredfold
excess of mercury or a thousandfold excess of bismuth.   More
electronegative elements do not influence the discharge or ionizatio
of silver.

Mercury

Hg^{2+} ions are reduced on a graphite electrode from nitrate
($\varphi_{1/2}$= -0.1 V), thiocyanate ($\varphi_{1/2}$= -0.25 V), chloride, citrate, and other
solutions.   The polarization curves for the stripping of the metal
have two maxima when nitric acid solutions are employed and one
well-defined maximum (φ_{max}= -0.1 V) in cyanide solutions.   The
optimal deposition potential for mercury is from -0.5 to -0.6 V.
The maximum anodic current is directly proportional to the concentration of the metal ions in solution.

Methods for the determination of mercury in ammonium and
potassium oxalate /71/, potassium thiocyanate /75/, and lead /76/
have been proposed.

*Determination of mercury and silver in ammonium and potassium
oxalate* /71/.   Graphite (type I) and mercurous sulfate electrodes
are employed.

First, a 1 g salt sample is dissolved in 20 ml 1 M potassium thiocyanate.   The solution is transferred to the electrolytic cell and
oxygen is removed by bubbling inert gas.   Electrolysis of the stirred
solution is carried out for 30 min at -1.2 V.   The stirring is stopped,
and the anodic polarization curve is recorded from -1.2 to -0.4 V.
The maximum current for the stripping of silver and mercury are
measured at potentials of approximately -0.6 and -0.5 V, respectively.
The concentration of metal ions is found from a known addition.

In the determination of $2 \cdot 10^{-6}\%$ silver and $4 \cdot 10^{-6}\%$ mercury, the
coefficient of variation is 12%.

Determination of silver and mercury in lead /76/.   Graphite
(type I) and mercurous sulfate electrodes are employed.

First, a 5 g pulverized lead sample is dissolved in 20 ml nitric
acid (1:3) in a quartz beaker.   The solution is evaporated to the
moist salt state, the volume is brought up to 50 ml with doubly
distilled water, 5 ml 1 N nitric acid and 5 ml 0.1 N versene are added.
The silver and mercury are extracted three times by 0.001% dithizone in carbon tetrachloride in 2 ml portions.   The organic extracts
are collected in a separating funnel and washed with doubly distilled
water.   Then, 5 ml 1 M potassium cyanide and 0.1 ml 0.1 N solution
versene are added to the washed extract and the silver is reextracted.   The mercury is reextracted by a solution 40% in potassium
bromide and 0.1 N in versene.   The reextracts are filtered through
a No. 1 Schott filter into polarography cups.   The prepared solutions
are boiled for one to two minutes.   Each of the resulting solutions
is placed in an electrolytic cell and oxygen is removed by bubbling
nitrogen.   Electrolysis of the stirred solution containing silver is

carried out at -0.8 V and the solution containing mercury at -0.9 V
for twenty minutes. The stirring is stopped, and the anodic polarization curves are recorded. The concentration of metal ions is
found from a known addition.

The coefficient of variation in the determination of $4 \cdot 10^{-8}\%$ silve
and $5 \cdot 10^{-9}\%$ mercury is 20%.

Determination of mercury and indium in zinc.* A graphite (type
and saturated calomel electrodes are employed.

First, an 0.1—1 g zinc sample is dissolved in 10—20 ml hydrochloric acid (1:1) and the solution evaporated to dryness. The dry
residue is dissolved in 15—20 ml 1.5—2 M potassium thiocyanate,
the oxygen being removed by bubbling inert gas for 10—15 min.
Electrolysis of the stirred solution is carried out for five to ten
minutes at -1.1 V. The stirring is stopped and the anodic polarization curve recorded after 30 sec. The maximum stripping curre
for indium is at -0.75 V and for mercury at -0.15 V. The concentration of metal ions is found from a known addition.

In the determination of $5 \cdot 10^{-6}\%$ mercury and indium, the coefficient of variation is 10—12%.

The determination of indium is not disturbed by a hundredfold
excess of lead or thousandfold excess of iron.

Copper

Copper ions are reduced on a graphite electrode to the metal
from acidic, neutral and alkaline solutions. The cathodic polarization half-wave potential and the anodic polarization curve maximum are strongly dependent on the nature of the complex ions
formed in the solution. Accordingly, the potential of the electrode
during the concentrating of copper must have a negative value.
Thus, copper is deposited from 1 M potassium nitrate at -0.6 V,
from a slightly alkaline sodium tartrate solution (pH = 9) at -0.8 V,
from a solution 0.1 M in potassium tartrate and 1 M in potassium
hydroxide at -1.6 V, and from a solution 1 M in potassium hydroxide and 1 M in sodium thiocyanate at -1.3 V.

The determination of copper is usually disturbed by antimony
and bismuth. The anodic currents of these elements are successfully separated by the choice of supporting electrolyte and electrochemical deposition potential. Thus, the potential difference of the
anodic current maxima of antimony and copper is adequate when

---

* The technique was proposed by E.Ya.Neiman and G.M.Dolgopolova.

an 8 N solution of hydrochloric acid with 20% tartaric acid is used as the supporting electrolyte, and commensurate amounts of bismuth do not disturb the determination of copper in an 0.5 N hydrochloric acid solution with 0.2% salicyl aldoxime, 20% citric acid, and $1 \cdot 10^{-4}$ N mercury (II) nitrate.   In this case, cadmium, the sum of the lead and tin, copper, antimony, and bismuth may be determined simultaneously (Figure 15).   Bismuth does not interfere with the determination of copper in a supporting electrolyte solution 1 N in potassium thiocyanate and 0.1 N in nitric acid when the ratio Cu:Bi is between 0.005:1 and 20:1.

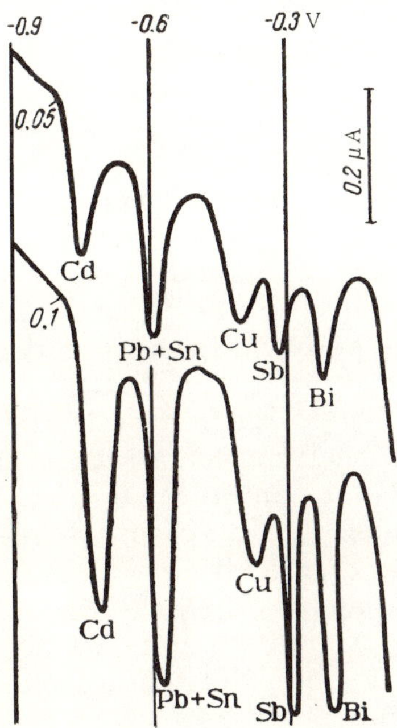

FIGURE 15. Polarization curves for the stripping ($v = 0.33$ V/min) of Cd, Pb, Sn, Cu, Sb, and Bi deposited at $\varphi_{el} =$ -0.9 V, $\tau_1 = 5$ min, from a solution 0.5 N in HCl, 0.2% in salicyl aldoxime, 20% in citric acid, and $1 \cdot 10^{-4}$ in Hg(NO$_3$)$_2$ and equal concentrations of the metal ions (numbers on the curves: μg/ml).

When copper is deposited together with silver, lead, iron, cobalt, and cadmium, distortions are found in the anodic polarization curves due to the formation of solid solutions or intermetallic compounds on the electrode.   The difficulties resulting from these distortions are eliminated either by choosing a deposition potential in the range in which discharge of the interfering ions does not occur or by adding mercury (II) nitrate to the solution being analyzed in the ratio $[\mathrm{Me}^{n+}]:[\mathrm{Hg}^{2+}] = 1:1000$.

Techniques for determining copper in cadmium and zinc sulfate, ammonium and potassium oxalates, hydrochloric acid, and metallic zinc using graphite (type I) and saturated calomel electrodes have been described.

*Determination of copper in hydrochloric acid.** First, 40 ml of the hydrochloric acid sample to be analyzed is evaporated down to 1 ml, 0.5 ml of a solution $5 \cdot 10^{-3}$ M in mercury (II) nitrate and 1 N in nitric acid are added, and the solution is diluted with water to 20 ml. The solution is transferred to an electrolytic cell and oxygen is removed by bubbling inert gas. Electrolysis of the stirred solution is carried out for 30 min at a graphite electrode potential of -0.6 V. The stirring is stopped, and after the solution settles, the anodic polarization curve is recorded from -0.6 to +0.1 V. The maximum stripping current for copper is measured. The concentration of metal ions is found from a known addition.

In the determination of $5 \cdot 10^{-8}\%$ copper, the coefficient of variation is 20%.

The determination is disturbed by commensurate amounts of bismuth.

*Determination of copper in cadmium and zinc sulfate.** First, a 6 g sulfate sample is placed in a heat-resistant beaker, moistened with water, and 20 ml hydrochloric acid is added. The mixture is heated and gradually evaporated to a syrupy state and 2—2.5 ml concentrated nitric acid is added. The solution is evaporated and another 1 ml nitric acid is added. The mass is cooled and dissolved in water, 0.5 ml of a solution $1 \cdot 10^{-3}$ M in mercury (II) nitrate and 1 N in nitric acid is introduced, and the solution evaporated and diluted to 30 ml. The solution is transferred to an electrolytic cell and oxygen removed by bubbling inert gas. Electrolysis of the stirred solution is carried out for 15 min at a graphite electrode potential of -0.55 V. The stirring is stopped and the polarization curve is recorded after 20—30 sec. Copper stripping occurs in the -0.2 to -0.1 V potential range. The copper ion concentration is found from a known addition.

In the determination of $3 \cdot 10^{-6}\%$ copper, the coefficient of variation is 20%.

For the determination of lead, cadmium, and copper in hydrochloric acid, ammonium and potassium oxalate, and copper and lead in zinc see the section below on lead.

The determination of copper, lead, and cadmium in zinc is described in the section below on cadmium.

---

* The technique was proposed by Kh.Z.Brainina and V.B.Belyavskaya.

Antimony

Antimony (III) ions are reduced on a graphite electrode to the elementary state. The half-wave potential of the cathodic polarogram recorded with a supporting electrolyte solution of 1 M hydrochloric acid is about -0.25 V and with a supporting electrolyte solution of 1 M ammonium acetate, -0.3 V. The amount of metal deposited during electrolysis from 1 M hydrochloric acid is independent of the electrode potential from -0.3 to -0.8 V and decreases abruptly if the cathodic process is carried out at a more negative potential. The reason for this anomaly evidently lies in the dislodgment of antimony anions from the electric double layer when the surface of the electrode is negatively charged /77, 78/.

The anodic polarization curves of antimony are well defined and have one current maximum directly proportional to the antimony ion concentration and the duration of electrodeposition. The potentials of the maxima are approximately -0.05 V (1 N HCl) and -0.25 V (1 N NH_4CH_3COO).

Copper and bismuth interfere with the determination of antimony. Commensurate amounts of antimony and copper can be determined using a solution 8 N in hydrochloric acid and 20% in tartaric acid serving as the supporting electrolyte. Commensurate amounts of antimony and bismuth may be determined and a solution 0.5 N in hydrochloric acid and 0.2% in salicyl aldoxime is employed as the supporting electrolyte.

The anodic currents of antimony, bismuth, and copper are better separated the less metal is deposited on the electrode in the concentrating step and the smaller the potential scanning rate in the stripping step. The combined deposition of antimony with lead, tin, and silver is accompanied by distortions in the anodic polarization curves due to the formation of intermetallic compounds on the electrode. These distortions may be prevented either by choosing a sufficiently positive deposition potential or by adding mercury (II) nitrate in the amount $[Me^{n+}]:[Hg^{2+}] = 1:1000$ to the solution.

Techniques for the determination of antimony in citric acid, tartaric acid, lead, tin, cadmium, and copper with the use of graphite (type I) and saturated calomel electrodes have been proposed.

*Determination of antimony in lead.** First, an 0.2 g lead sample is dissolved in 5 ml nitric acid (1:9) with gentle heating and the resulting solution is evaporated to dryness. The residue is dissolved in 5 ml water and the solution is slowly poured into 10 ml concentrated hydrochloric acid. The solution is transferred to an

---

* The technique was proposed by Kh.Z.Brainina and N.K.Kiva.

electrolytic cell, an inert gas is passed through a washing trap containing 6 N hydrochloric acid and then through the cell, and 0.2 ml of a solution containing 0.3 mg of Hg (II) is introduced. After 15 min the electrolysis of the stirred solution is carried out for 10 min at -0.60 V and the anodic curve is recorded. The stripping current of antimony is observed from -0.30 to -0.20 V.

The antimony ion concentration is found from a known addition.

A standard solution of antimony (III) is prepared by dissolving 115 mg Sb_2O_3 in 100 ml hydrochloric acid (1:2). A solution containing 2.5 μg Sb (III) in 1 ml is prepared by successive dilution of the standard solution with 1 N hydrochloric acid.

In the determination of $5 \cdot 10^{-5}\%$ antimony, the coefficient of variation is 15%.

*Determination of antimony in cadmium.** Graphite (type I) and saturated calomel electrodes are used. First, 5 ml concentrated hydrochloric acid and 1 ml nitric acid are placed in a platinum crucible and 1 g cadmium shavings is gradually added. After dissolution of the cadmium the volume of the solution is diluted to 15 ml with water. The resulting solution is placed in the electrolytic cell, 0.2 of a solution containing 1 mg of Hg (II) in 1 ml is added, and oxygen is removed by bubbling nitrogen for 10 min. The salt bridge is filled with 2 M hydrochloric acid. Electrolysis of the stirred solution is carried out at from -0.55 to 0.6 V. Then the system is maintained for 1 min at -0.35 V to dissolve the cadmium. The stirring is stopped and the anodic polarization curve is recorded. Antimony stripping is observed from -0.2 to -0.5 V. The antimony ion concentration is found from a known addition.

The standard solution is prepared as indicated in the previous technique.

The coefficient of variation in the determination of $5 \cdot 10^{-5}\%$ antimony in cadmium is 15%.

*Determination of antimony in tin.** First, 1 g tin shavings are placed in a platinum dish and dissolved with gentle heating in 10 ml concentrated hydrochloric acid. The resulting solution is diluted to a volume of 15 ml with distilled water. The solution is transferred to an electrolytic cell and oxygen is removed by bubbling nitrogen for 10 min. The bridge and the intermediate vessel are filled with 1 M hydrochloric acid. Electrolysis of the stirred solution is carried out for 15 min at -0.6 V. The system is maintained at -0.4 V until the tin is dissolved, the stirring is stopped, and without disconnecting the external circuit, the electrode is transferred to a solution 1 M in hydrochloric acid and 10^{-4} M in mercury (II) nitrate.

---

* The technique was proposed by E.Ya.Neiman and L.N.Trukhacheva.

The electrolysis is carried out for 1 min at -0.4 V and then the anodic dissolution polarization curve is recorded. Antimony stripping occurs from -0.2 to 0 V. The polarization curves and calibration graph are given in Figures 16 and 17. The figures show that under certain conditions tin does not interfere with the determination of antimony. The concentration of antimony is found from a known addition.

The standard solution is prepared as described above.

In the determination of $5 \cdot 10^{-5}\%$ antimony, the coefficient of variation is 20%.

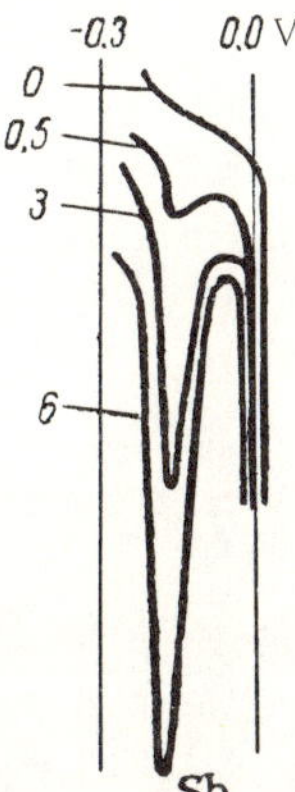

FIGURE 16. Polarization curves for the electrochemical dissolution ($v = 1$ V/min) of antimony deposited at $\varphi_{el} = -0.55$ V, $\tau_1 = 15$, min from solutions with different Sb (III) concentrations (numbers on the curves: $\times 10^{-7}$ g · ion/liter).

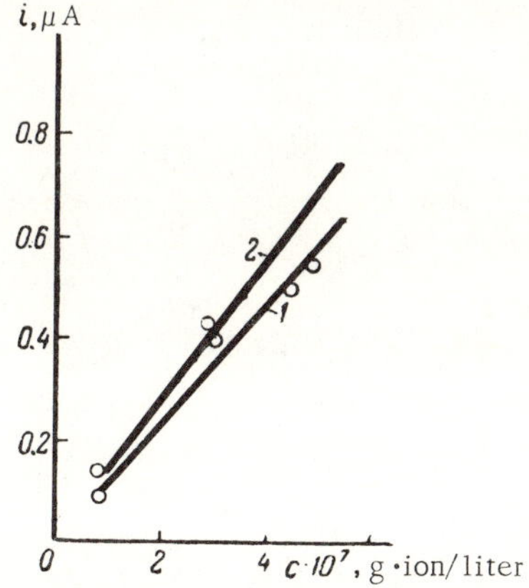

FIGURE 17. Dependence of the maximum stripping current of antimony (conditions of Figure 16) on the concentration of antimony ions in different solutions:

1) 8 M HCl solution with an Sn (II) concentration of 0.3 g · ion/liter;
2) 6 M HCl solution with a Cd (II) concentration of 0.3 g · ion/liter.

*Determination of antimony in copper.** First, an 0.5—1 g copper sample is dissolved in 20 ml nitric acid (1:1) and the solution is boiled to remove the nitrogen oxides and diluted to 150—200 ml with water. Then, 2 ml of a ferric chloride solution containing 10 mg Fe (III) per ml are introduced and ammonium hydroxide is added until the formation of a dark blue copper ammoniate is noted. The solution is heated to 75—80°C and held at this temperature for 10 to 12 min. The residue is collected by filtration and washed two or three times with hot 3% ammonium hydroxide. The residue is washed with hot water into the beaker in which the precipitation took place and dissolved in 30 ml hot hydrochloric acid (1:1). Reprecipitation is repeated three times. Then, the residue is dissolved on the filter in 20—25 ml 1—3 N hydrochloric acid, 1 g hydrazine chloride is added to the solution, and is then evaporated for 2 min. The solution is cooled, 1.5—2 g citric acid is added, and the solution is transferred to the electrolytic cell. Then, 60—100 μg of mercury (II) are added and oxygen is removed by bubbling inert gas. Electrolysis is carried out for 5 min at -0.5 V and then the anodic polarization curve is recorded. The maximum stripping current of antimony is observed at -0.3 V. The antimony ion concentration is found from a known addition.

In the determination of $5 \cdot 10^{-6}$% antimony, the coefficient of variation is 20%.

The determination is not disturbed by a twentyfold excess of bismuth.

The determination of antimony in citric and tartaric acids in the presence of tin is described in the section on tin and the determination of antimony and bismuth in lead, in the section on bismuth.

Bismuth

Bismuth ions are reduced on graphite electrodes to the metal which dissolves upon anodic polarization of the electrode. The observed polarization curve has a very sharp maximum proportional to the bismuth ion concentration and to the deposition time. The optimal potential for bismuth deposition from a solution of 0.1—1.5 hydrochoric acid, which is the supporting electrolyte, is from 0.4 to 0.5 V. Metal stripping occurs from -0.3 to -0.05 V. The determination of bismuth is not disturbed by more electronegative elements (lead, cadmium, thallium, etc.). The anodic currents of coppe

---

* The technique was proposed by E.Ya.Neiman and L.N.Trukhacheva.

antimony, and bismuth are successfully separated when a solution 0.5 N in hydrochloric acid, $1 \cdot 10^{-4}$ M in mercuric nitrate, 20% citric acid, and 0.2% salicyl aldoxime serves as the supporting electrolyte (see Figure 15). Copper does not interfere with the determination of bismuth in a solution 1 N in KSCN and 0.1 N in HNO_3 for ratios of [Cu]:[Bi] up to 20:1.

Techniques for the determination of bismuth in tin, citric acid, and tartaric acid, copper, and lead employing graphite (type I) and saturated calomel electrodes have been proposed.

Determination of bismuth in hydrochloric acid. First, 40 ml of the acid to be analyzed are evaporated to 1 ml, 0.5 ml of a solution $5 \cdot 10^{-3}$ M in mercury (II) nitrate and 1 N in nitric acid are added, and the solution is diluted to 20 ml with water. The solution is transferred to the electrolytic cell and oxygen is removed by bubbling inert gas. Electrolysis of the stirred solution is carried out for 15 min at a graphite electrode potential of -0.4 V. The stirring is stopped and the anodic polarization curve is recorded after the solution becomes still. The maximum stripping current for bismuth is measured. The bismuth ion concentration is found from a known addition.

In the determination of $2.5 \cdot 10^{-8}$% bismuth, the coefficient of variation is 20%.

Determination of bismuth in tin /80/. Preparation of the graphite electrode. First, 20 ml 2 M hydrochloric acid is placed in the electrolytic cell and the salt bridge is filled with the same solution. Then, 1 ml of a solution $1 \cdot 10^{-3}$ M in mercury (II) nitrate and 1 M in nitric acid is added to the electrolytic cell. Electrolysis of the stirred solution is carried out for 5 min at a graphite electrode potential of -0.3 V. Finally, the electrode is removed from the solution under the same voltage and washed with distilled water.

Determination procedure. First, 0.5 g tin shavings are dissolved in 3 ml concentrated hydrochloric acid with heating and several drops of nitric acid are added. The volume of the solution is brought up to 15 ml with distilled water, the solution is placed in the electrolytic cell, and an inert gas is passed. The salt bridge is filled with 1.5 M hydrochloric acid. Electrodeposition from the stirred solution is carried out for 10 min at -0.4 V. The electrode potential is decreased to -0.3 V, which is maintained until complete stripping of the tin. The stirring is stopped and, after 30 sec, the anodic poraization curve is recorded at a potential scanning rate of 400 mV/min. Bismuth stripping is observed from -0.15 to 0.05 V. The concentration of bismuth is found from a known addition.

In the determination of $3 \cdot 10^{-4}$% bismuth, the coefficient of variation is 10%.

*Determination of bismuth in copper.** First, an 0.5—1 g copper sample is dissolved in 20 ml nitric acid (1:1). The solution is boiled for 2—3 min to remove nitrogen oxides and diluted to 150—200 ml with water. Then, 2 ml of a lanthanum chloride solution containing 10 mg La per ml are introduced and ammonium hydroxide is added until the formation of a dark blue copper ammoniate is observed. The solution is heated to 75—80°C and maintained at this temperature for 10—12 min. The residue is collected using an average density filter and washed on the filter two or three times with hot 3% ammonium hydroxide. The residue is washed from the filter with hot water into the beaker in which the precipitation occurred and dissolved in 30 ml hot hydrochloric acid (1:1). Reprecipitation is repeated three times. After the third precipitation, the residue is dissolved on the filter in 20—25 ml 1—3 N hydrochloric acid.

The solution is cooled, 1.5—2 g citric acid added, and the solution is transferred to the electrolytic cell. Then, 60—100 μg mercury (II) is added and oxygen is removed by bubbling argon for 7 to 10 min. Electrolysis of the stirred solution is carried out for 5 min at -0.45 V. The stirring is stopped and the solution is allowed to settle for 30 sec. Next, the anodic polarization curve is recorded. The maximum anodic current for bismuth is found at -0.15 V. The concentration of bismuth is found from a known addition.

In the determination of $5 \cdot 10^{-6}\%$ bismuth, the coefficient of variation is 20%.

The determination is not disturbed by a twentyfold excess of antimony.

*Determination of antimony and bismuth in lead.** First, an 0.2—0.5 g lead sample is dissolved in 15—20 ml nitric acid (1:1), the solution is evaporated to dryness and the dry residue is dissolved in 20 ml hydrochloric acid (1:1). The solution is again evaporated to dryness and the residue dissolved in 20 ml hydrochloric acid. Evaporation with hydrochloric acid is repeated another three times. The dry residue is dissolved in 20 ml 8 N hydrochloric acid and 1.5—2 g citric acid and 100 μg mercury (II) are added. The solution is transferred to the electrolytic cell and oxygen is removed by bubbling inert gas for 10 to 12 min. Electrolysis of the stirred solution is carried out for 2—5 min at -0.5 V. The stirring is stopped, the solution is allowed to settle for 30 sec, and the anodic polarization curve is recorded.

Antimony and bismuth stripping is observed at from -0.35 to 0.10 V. The concentration of metal ions is found from a known addition.

---

* The technique was proposed by E.Ya.Neiman and L.N.Trukhacheva.

The determination is not disturbed by a twentyfold excess of copper or a two-hundredfold excess of tin.

In the determination of $1 \cdot 10^{-5}\%$ antimony and bismuth, the coefficient of variation is 15%.

The simultaneous determination of lead, cadmium, and total bismuth and copper in hydrochloric, tartaric, and citric acids is described in the section on lead.

Lead

The half-wave potentials of the cathodic polarograms of lead are -0.6 and -0.7 V in the reduction with a supporting electrolyte solution of 0.1 N hydrochloric acid (or 0.1 N ammonium chloride solution) and 20% citric acid.   The potentials of the anodic polarization curve maxima are close to from -0.55 to -0.5 V.   The optimal conditions for deposition exist at -1.0 V.   Well-defined stripping curves are obtained in the electrochemical deposition of lead from solutions with a Pb^{2+} concentration of $1 \cdot 10^{-8}$ g·ion/liter for 10 min.

In the presence of copper or antimony, which are usually deposited on the electrode together with lead, distortions are observed in the anodic polarograms.   Introduction of a quantity of mercury (II) nitrate a thousand times greater than these metals eliminates this phenomenon.

The determination of lead in 1 N hydrochloric acid is not disturbed by a tenfold excess of chromium, nickel, or titanium, a hundredfold excess of iron, or commensurate amounts of tin.

Lead may be determined in the presence of a tenfold amount of tin in supporting electrolyte solutions 1 M in ascorbic acid and nitric acids and 2 M in phosphoric acid and perchloric acids.

Techniques have been proposed for the determination of lead in hydrochloric acid, ammonium and potassium oxalates /71/, nitric acid /81/, potassium nitrate /82/, citric and tartaric acid, copper and its alloys employing graphite (type I) and saturated colomel electrodes in hydrochloric acid.

Determination of lead, cadmium, and copper in hydrochloric acid.
First, 40 ml of the hydrochloric acid sample to be analyzed are evaporated down to 1 ml, 0.5 ml of a solution $5 \cdot 10^{-3}$ M in mercury (II) nitrate and 1 M in nitric acid is added, and the solution diluted to 20 ml with water.   The solution is transferred to the electrolytic cell and oxygen is expelled by bubbling inert gas.   Electrolysis of the stirred solution is carried out for 30 min at a graphite electrode potential of -1.2 V.   The stirring is stopped and the anodic polarization curve is recorded.   The maximum stripping currents of

cadmium, lead, and copper are measured at -0.7—0.8 V, -0.5— -0.6 V and 0—0.1 V, respectively. The concentrations of the metals are found from a known addition.

In the determination of $1 \cdot 10^{-7}\%$ lead, $5 \cdot 10^{-8}\%$ cadmium, and $5 \cdot 10^{-8}\%$ copper, the coefficient of variation is 20%.

The determination of copper is disturbed by bismuth.

Determination of lead, cadmium, and copper in ammonium and potassium oxalates /71/. First, a 1 g salt sample is dissolved in 20 ml 1.5% potassium citrate. The solution is transferred to the electrolytic cell, oxygen is removed by bubbling inert gas, and 0.5 ml of a solution $5 \cdot 10^{-3}$ M in mercury (II) nitrate and 0.1 M in nitric acid are introduced. Electrolysis of the stirred solution is carried out for 30 min at a graphite electrode potential of -1.2 V. The stirring is stopped and the anodic polarization curve is recorded. The maximum oxidation currents of the metals are measured at potentials of about -0.7 V (cadmium), -0.55 V (lead), and -0.2 V (copper). The metal ion concentrations are found from a known addition.

In the determination of $4 \cdot 10^{-6}\%$ lead, $2 \cdot 10^{-6}\%$ cadmium, and $2 \cdot 10^{-6}\%$ copper in ammonium and potassium oxalates, the coefficient of variation is 15%.

*Determination of lead in citric and tartaric acid.** First, 4 g of the acid sample to be analyzed are dissolved in water and the volume of the solution is diluted up to 20 ml. The solution is transferred to the electrolytic cell, 0.2 ml of a mercury solution containing 1 mg Hg (II) per ml are added and oxygen is expelled by bubbling inert gas. Electrolysis of the stirred solution is carried out for 5 min at -1.2 V. The stirring is stopped, the voltage is decreased to -0.8 V, and, after 15—20 sec, the polarization curve is recorded from -0.8—0 V. The maximum stripping current of lead is measured at a potential of approximately -0.5 V.

The concentration of lead is found from a known addition.

In the determination of $1 \cdot 10^{-5}\%$ lead in citric and tartaric acids, the coefficient of variation is 15%. Tin and antimony do not interfere with the determination.

*Determination of lead and copper in zinc.*** Under heating, 0.1—1 g zinc is dissolved in 10—20 ml hydrochloric acid (1:1), several drops of nitric acid are added, and the solution is evaporated two or three times with hydrochloric acid to dryness. The dry residue is dissolved in 15—20 ml 0.1 M acetate buffer solution with gentle heating without boiling. The solution is cooled, 100 μg mercury (II) is added, the solution is placed in the electrolytic cell, and

---

* The technique was proposed by Kh. Z. Brainina and E. M. Roizenblat.
** The technique was proposed by L. N. Trukhacheva and G. I. Dolgopolova.

oxygen is removed by bubbling inert gas.   Electrolysis of the
stirred solution is carried out for 3—5 min at -0.8 V.   The stirring
is stopped and the anodic polarization curve is recorded after 30 sec.
The maximum stripping current of lead is observed at from -0.5
to -0.45 V and of copper, at from -0.3 to -0.25 V.   The metal ion
concentrations are found from a known addition.

In the determination of $5 \cdot 10^{-5}\%$ copper and lead, the coefficient
of variation is 10%.

*Determination of lead in copper alloys.** First, 0.1—1 g copper
alloy is dissolved in 20 ml nitric acid (1:1).   The solution is boiled
for 2—3 min to remove nitrogen oxides and diluted with water to a
volume of 150—200 ml.   If there is an insufficient amount of iron in
the sample, 20 mg of Fe (III) are introduced into the solution.   In the
analysis of alloys containing sufficient iron, this step is omitted.
An ammonia solution is added until the formation of a dark blue copper ammoniacal complex is observed, and the solution is heated to
75—80°C.   The residue is collected on a filter of average density and
washed three or four times on the filter with a hot ammonia solution.
The precipitate is washed from the filter with hot water into the
beaker in which the precipitation occurred and dissolved in 30 ml
hot hydrochloric acid (1:1).   For more complete separation of copper, a triple reprecipitation is performed.   Then, the solution is
evaporated and transferred to a 50 ml volumetric flask and a medium
is created 3N in hydrochloric acid.

Then, 10—15 ml 0.1 M acetate buffer solution is placed in the
electrolytic cell and oxygen is removed by bubbling inert gas for
10—12 min.   Next, 1—5 ml of the solution to be analyzed are transferred to the electrolytic cell, 1—2 g hydrazine chloride and $40-60 \mu g$
mercury (II) are added, and inert gas is passed again.   Electrolysis
of the stirred solution is carried out for 3 min at -1.0 V; then the
stirring is stopped, the solution is allowed to settle for 25—30 sec,
and the anodic polarization curve is recorded.   The stripping current maximum of lead is observed at from -0.5 to -0.45 V.   The
lead ion concentration is found from a known addition.

In the determination of $1 \cdot 10^{-4}\%$ lead, the coefficient of variation
is 10%.

The determination of lead, copper, and cadmium in zinc and lead
is described in the section on cadmium.

Tin

Tin ions are reduced to the metal in the presence of chloride ions
at potentials more negative than -0.6 V.   The limiting cathodic

---

* The technique was proposed by L.N.Trukhacheva and G.I.Dolgopolova.

current of tin in 1 N hydrochloric acid is found at -0.8 V. The stripping current maximum is independent of the electrode potentia in the cathodic cycle from -1.0 to -1.2 V. The potential of the anodic polarization curve maximum is approximately -0.5 V in the oxidation of tin in a supporting electrolyte solution of 1 M ammonium chloride. Commensurate amounts of lead do not interfere with the determination of tin in 0.5 M oxalic acid or 8 N hydrochloric acid.

The effect of silver and antimony on the polarization curves for the stripping of tin has been detected, but is absent when the solutio contains Hg^2 ions in the ratio $[Me^{n+}]:[Hg^{2+}] = 1:1000$.

Techniques for the determination of tin in citric and tartaric acids employing graphite (type I) and saturated calomel electrodes have been proposed.

*Determination of tin and antimony in citric and tartaric acids.** First, 4 g of the acid sample are dissolved in 15 ml concentrated hydrochloric acid and the volume of the solution is diluted up to 20 r with water. The solution is transferred to the electrolytic cell, oxygen is expelled by bubbling inert gas, and 0.2 mg of mercury (II) is added. Electrolysis of the stirred solution is carried out for 10 min at -0.9 V. The stirring is stopped, the voltage decreased to -0.8 V, and after 15—20 sec, the anodic polarization curve is record ed at from -0.8 to -0.2 V. The maximum current for tin stripping is measured at -0.65 V. The anodic polarization curves of tin are shown in Figure 18. A fivefold excess of lead does not interfere with the determination.

For the determination of antimony, the same solution is electrolyzed for 5 min at -0.5 V. The anodic polarization curve is recorded at from -0.5 to -0.2 V. The maximum current for antimony stripping is measured at approximately -0.3 V. The polarization curves and the dependence observed under these conditions between the maximum current of the electrochemical oxidation of antimony and the concentration of antimony ions in the solution are presented in Figures 19 and 20.

The metal concentrations are found from known additions.

The standard solution of tin is then diluted by 3 N hydrochloric acid. The standard solution of antimony (III) is prepared by dissolving Sb_2O_3 in hydrochloric acid (1:2) and subsequently diluting with hydrochloric acid.

In the determination of $1 \cdot 10^{-5}\%$ tin and antimony, the coefficient of variation is 20%.

---

* The technique was proposed by E. M. Roizenblat and Kh. Z. Brainina.

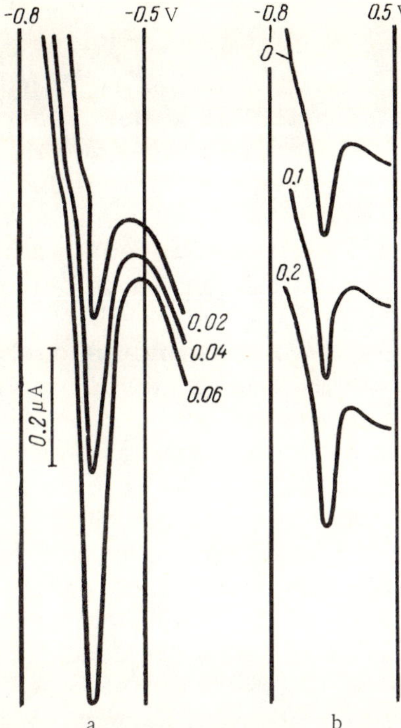

FIGURE 18. Polarization curves for the stripping ($v = 1$ V/min) of tin deposited at $\varphi_{el} = 1.2$ V, $\tau_1 = 5$ min from various solutions:

a) with different concentrations of Sn (II) (numbers on the curves, μg/ml); b) with an Sn (II) concentration of 0.02 μg/ml and different concentrations of Pb (II) (numbers on the curves, μg/ml).

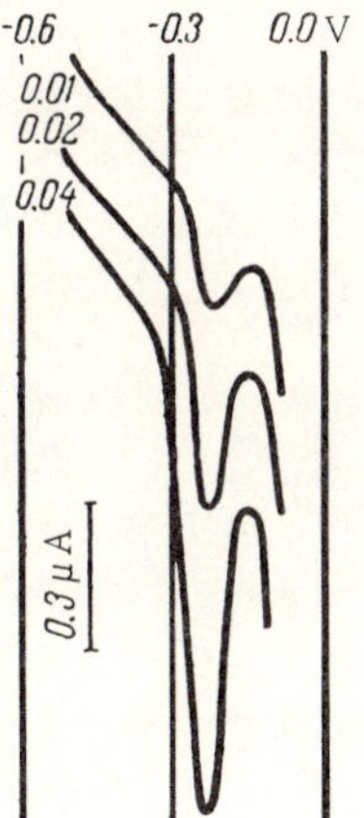

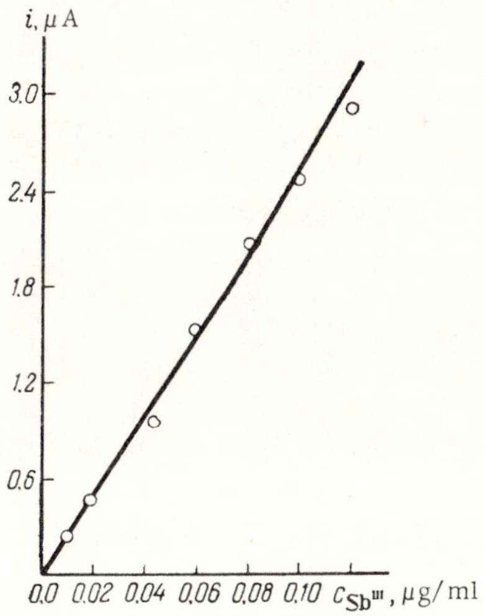

FIGURE 19. Polarization curves for the stripping ($v = 1$ V/min) of antimony deposited at $\varphi_{el} = -0.5$ V, $\tau_1 = 5$ min from solutions 8 N in HCl 20% in citric acid, and containing different amounts of Sb (III) (numbers on the curves, μg/ml).

FIGURE 20. Dependence of the maximum stripping current (conditions of Figure 19) of antimony on the concentration of antimony ions in the solution.

Cadmium

The half-wave potential of the cadmium cathodic polarogram in 0.1 N hydrochloric acid equals -0.8 V, the potential of the tin anodic polarization curve maximum is about -0.75 V, and the potential necessary for the deposition of cadmium from very dilute solutions is -1.0— -1.2 V.

A strong effect of copper and antimony deposited together with cadmium on the nature of the anodic polarization curves is observed due to the formation of intermetallic compounds on the electrode. The latter do not arise when mercury is deposited simultaneously on the electrode employing a thousandfold excess of mercury ions.

Methods of determining cadmium, lead, and copper in zinc, cadmium and lead in copper with the use of graphite (type I) and saturated calomel electrodes are described.

*Determination of cadmium, lead, and copper in zinc.** Under heating, an 0.5 g zinc sample is dissolved in 5 ml hydrochloric acid (1:1), several drops of concentrated nitric acid are added, and the solution is evaporated two or three times with hydrochloric acid to dryness. The salts obtained are dissolved in 4—5 ml hydrochloric acid (1:1) and diluted with water. The cooled solution is transferred to a 50 ml volumetric flask, 1 ml of a solution containing $300 \mu g$ Hg (II) per ml of 1 N nitric acid is added, and the solution is diluted to 50 ml with water. Then, 20 ml of the resulting solution is transferred to the electrolytic cell and oxygen is removed by bubbling inert gas for 10—15 min. Electrolysis of the stirred solution is carried out for 10—20 min at -0.95 V. The stirring is stopped, the solution is allowed to settle, and the anodic polarization curve is recorded at from -0.95 to +0.1 V. The maximum stripping current for cadmium is measured at -0.75 V, for lead at -0.55 V, and for copper at -0.2 V. The metal concentrations are found from a known addition.

In the determination of $5 \cdot 10^{-6}\%$ copper, lead, and cadmium, the coefficient of variation does not exceed 20%.

*Determination of cadmium and lead in copper.*** First, a 1 g copper sample is dissolved in a quartz beaker in 10—15 ml of nitric acid (1:1), and the solution is boiled for 2—3 min to remove nitrogen oxides. The solution is cooled and diluted by doubly distilled water to a volume of 60—70 ml and 0.5 ml concentrated sulfuric acid is added. Electrolysis is carried out with a platinum grid cathode for

---

  * The technique was proposed by Kh. Z. Brainina, N. D. Fedorova, and V. M. Vdovina.
** The technique was proposed by E. Ya. Neiman, G. M. Dolgopolova, and L. N. Trukhacheva.

1—1.5 hr at 5 A and 2 V.   Then, the solution is evaporated in the beaker to the moist salt state, and the salts are dissolved in 10—15 ml 1 M potassium chloride.   The solution is transferred to the electrolysis cell, 500 μg of mercury (II) are added, and oxygen is removed by bubbling inert gas for 10—12 min.   Electrolysis of the stirred solution is carried out for 20 min at -1.1 V.   The stirring is stopped, the solution allowed to settle for 1 min, and the anodic polarization curve is recorded.   The potentials of the stripping current maxima of cadmium and lead are observed at from -0.7 to -0.6 V and from -0.5 to -0.45 V, respectively.   The concentration of the metal ions is found from a known addition.

In the determination of $1 \cdot 10^{-5}\%$ cadmium and lead, the coefficient of variation does not exceed 20%.

The duration of the analysis is three hours.

The determination of cadmium is not disturbed by a fairly large excess of lead and copper.   The determination of lead is not disturbed by a twenty-fivefold excess of cadmium and a two-hundredfold excess of copper.

Indium

Figures 21 and 22 illustrate the cathodic polarograms of indium in potassium chloride supporting electrolyte solutions and thiocyanate (pH = 5) and the anodic polarization curves for the dissolution of the metal previously deposited on the electrode.   The anodic polarization curves have the usual shape with a maximum current, increasing as the electrode potential is shifted in the negative direction to -1.3 V when the preliminary metal deposition is carried out from a potassium thiocyanate solution (Figure 21b), and to -1.2 V when the metal is deposited from a potassium chloride solution (Figure 22b).   Electrolysis at the more negative potential is accompanied by a decrease in the metal stripping current.   This dependence evidently arises as the result of a reorientation of indium complexes in the electrical double layer under the negative electrode surface charge during deposition, which causes a decrease in the electroreduction current of the metal ions /78/ and a decrease in the amount of indium deposited on the electrode (see equation (II. 1)).

For the determination of indium and mercury in zinc see the section on mercury.

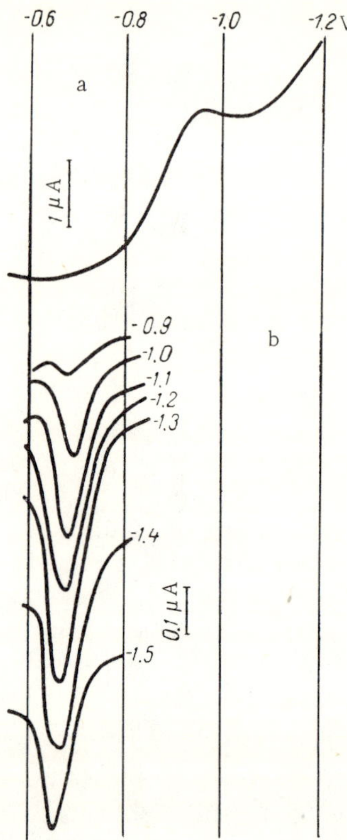

FIGURE 21. Polarization curves for indium in 1 M KSCN (pH = 5, v = 0.01 V/sec):

a) cathodic curve at an In (III) concentration of $5 \cdot 10^{-4}$ g · ion/liter; b) anodic curves for indium deposited at τ_1 = 2 min, an In (III) concentration of $4 \cdot 10^{-6}$ g · ion/liter, and various φ_{el} (numbers on the curves, V).

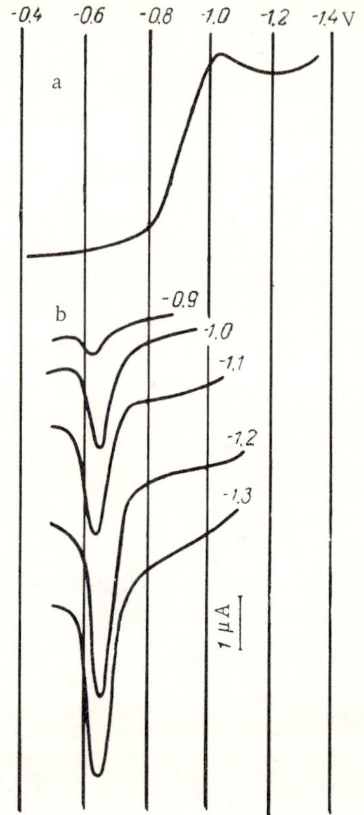

FIGURE 22. Polarization curves of indium in 0.25 M KCl (pH=5, v =0.017 V/sec):

a) cathodic curve at an In (III) concentration of $5 \cdot 10^{-4}$ g · ion/liter; b) anodic curves for indium deposited at an In (III) concentration of $4 \cdot 10^{-6}$ g · ion/liter, τ_1 = 1 min, different φ_{el} (numbers on the curves, V).

Thallium

Thallium ions are reduced on the graphite electrode from acid, neutral, and slightly alkaline solutions. The optimal potentials for electrodeposition of the element are -1.4 V from an ammonium-ammonia buffer solution and from -1.0 to -1.2 V from neutral and acidic perchlorate solutions. The half-wave potentials of the cathodic polarograms differ slightly from the potentials of the anodic curve maxima. This process is apparently reversible. The anodic polarization curves of thallium are pronounced. The maximum metal stripping current is proportional to the metal ion concentration in the solution and the deposition time.

Gallium

Gallium ions are reduced on the graphite electrode from potassium thiocyanate and nitrate solutions (pH = 5). The cathodic curves are fairly well defined and the half-wave potentials are found at about -1.4 and -1.2 V. The limiting current decreases when the gallium ion concentration in the solution is increased. There is no marked anodic current maximum in the anodic polarization curves of gallium. When the gallium ion concentration or the electrolysis time increases, two different current maxima appear on the curves. However, accurate measurements from these curves are difficult.

Cobalt

The discharge of cobalt ions occurs at potentials more negative than -1.0 V, nevertheless the cathodic polarograms are well defined (Figure 23). The anodic polarization curves of cobalt have a characteristic current maximum (Figure 24) directly proportional to the cobalt ion concentration (Figure 25). The element may be determined in supporting electrolyte solutions containing sulfate, thiocyanate, tartrate, and ammonium ions. The analogous curves after electrochemical deposition at the same potentials from thiocyanate and ammonia solutions yield nickel, while nickel is not deposited from strongly alkaline solutions containing tartrate or thiocyanate ions. Such solutions may therefore be employed to determine cobalt in the presence of considerable amounts of nickel. Usually, the determination of cobalt is not disturbed by commensurate amounts of mercury, silver, bismuth, lead, cadmium, as well as by a number of other elements.

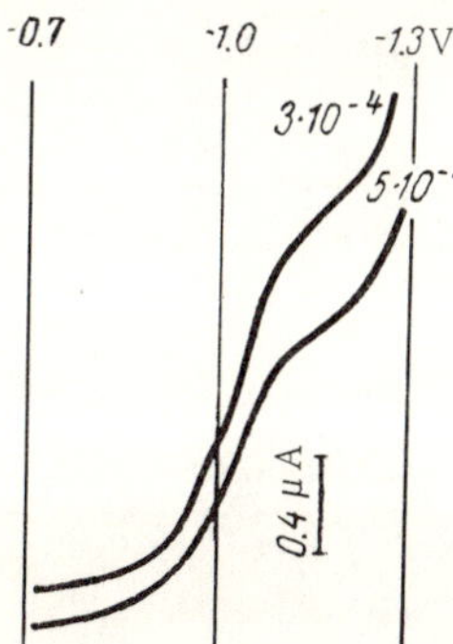

FIGURE 23. Cathodic polarization curves of cobalt in an 0.1 M K_2SO_4 supporting electrolyte solution at different Co^{2+} concentrations (numbers on the curves: g · ion/liter).

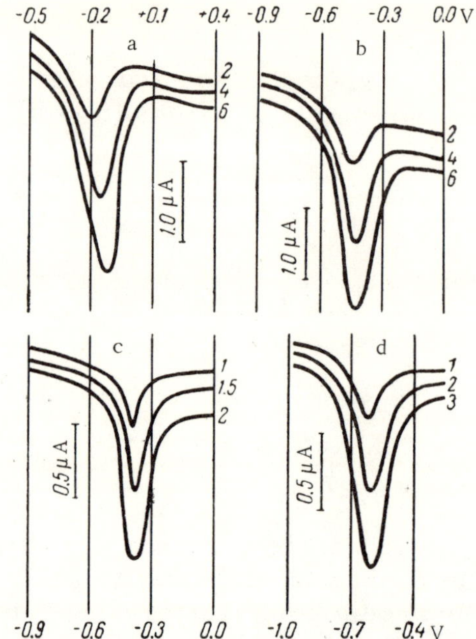

FIGURE 24. Polarization curves for the stripping of cobalt deposited at $\tau_1 = 5$ min from solutions with different Co^{2+} concentrations (numbers on the curves: $\times 10^{-6}$ g $\cdot$ ion/liter) under the conditions:

a) 0.1 M K_2SO_4 solution, $\varphi_{el} = -1.4$ V; b) 0.1 M NaSCN solution, $\varphi_{el} = -1.2$ V; c) 0.1 M NH_4OH solution, φ_{el} -1.6 V; d) solution 1 M in KOH and 0.1 M in potassium tartrate, φ_{el} -1.2V.

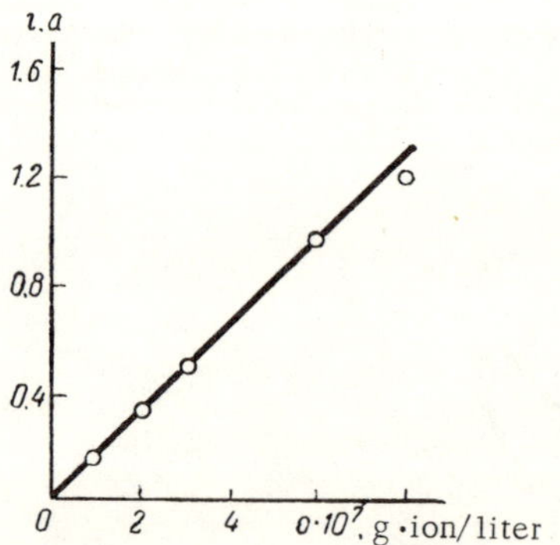

FIGURE 25. Dependence of the maximum anodic current on the cobalt ion concentration in 0.1 M NaSCN for $\tau_1 = 20$ min.

*Determination of cobalt in ammonium molybdate.** Graphite (type I) and saturated calomel electrodes are employed.

The sample is roasted for 4 hr in an open muffle furnace at $600°C$. Then 1 g of the calcined preparation is dissolved in 20 ml of a solution containing 100 g of sodium thiocyanate and 150 g of potassium hydroxide per liter. The resulting solution is transferred to the electrolytic cell and oxygen is removed by bubbling inert gas. The electrolysis of the stirred solution is carried out for 20 min at a graphite electrode potential of -1.5 V. The stirring is stopped and the anodic polarization curve recorded. The maximum stripping current for cobalt is measured and the concentration is found from a known addition.

In determining $1 \cdot 10^{-5}\%$ cobalt, the variation coefficient amounts to 10%.

The technique may be used for analyzing sodium molybdate, sodium tungstate, and ammonium.

Nickel

Fairly pronounced cathodic polarograms of nickel on the graphite electrode may be obtained only in the reduction of complex ions of the metal. The discharge of the aqua-ions are masked by the reduction current of hydrogen ions though this does not interfere with concentrating the element. The anodic polarization curves of nickel are well defined. The maximum current is proportional to the metal ion concentration in the solution and the deposition time. The dissolution process which is accompanied by the formation of aqua-ions occurs in the most positive potential range. When a solution 1 M in KNO_3 and 0.001 M in HNO_3 serves as the electrolyte, the potential of the anodic current maximum is about -0.1 V. The maximum oxidation current of nickel in an 0.1 M sodium thiocyanate solution is observed at -0.4 V, while in a solution 0.1 M in NH_4OH and 0.1 M in NH_4Cl, at -0.5 V.

The determination of nickel is not disturbed by commensurate amounts of mercury, silver, bismuth, lead, cadmium, as well as by a number of other elements. An exception is copper.

Iron

Iron ions are reduced to the metal on a graphite electrode from solutions containing complex ions of the element. The optimal

---

* The technique was proposed by E.M.Roizenblat and Kh.Z.Brainina.

deposition potentials from an 0.05 M sodium sulfosalicylate solution (pH = 5) and a solution 1 M in KOH and 0.001 M in sodium tartrate are -1.6 and -1.4 V. When the concentration of the hydroxyl ions in the solution is increased, the electrolysis potential should be more negative. In both cases well-defined polarization curves are obtained in the anodic polarization of the electrode (Figure 26). It is more convenient to use a solution 1 M in KOH and 0.01 M in sodium tartrate as the supporting electrolyte since hydrogen ions are not discharged in the process of concentrating iron and the results of the analysis are more readily reproducible.

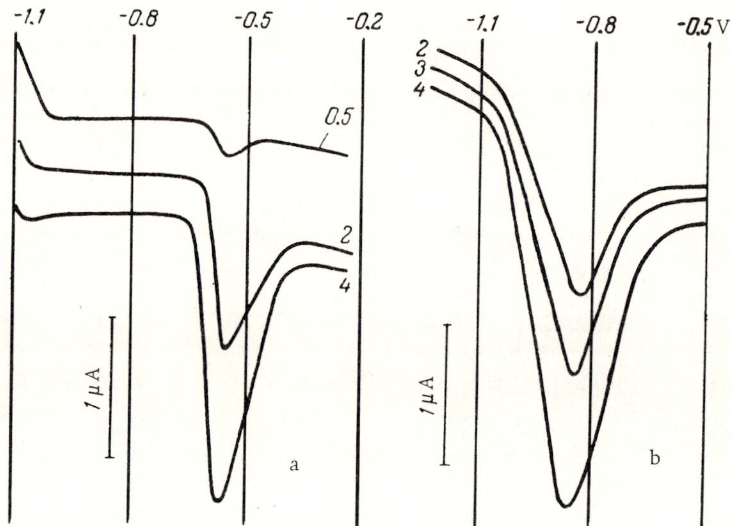

FIGURE 26. Stripping polarization curves of iron deposited from solutions with different Fe^{3+} concentrations (numbers on the curves: $\times 10^{-6}$ g · eq/liter) at $\tau_1 = 5$ min under the conditions:

a) a solution 0.05 M in sodium sulfosalicylate (pH=5) at $\varphi_{el} = -1.6$ V; b) a solution 1 M in KOH and 0.01 M in sodium tartrate, $\varphi_{el} = -1.4$ V.

Iron may also be deposited on the electrode from alkaline solutions (pH = 10) containing tartrate or citrate ions. However, a second current maximum then appears on the anodic polarization curves, probably due to the intermediate formation of $Fe(OH)_2$ films since a high concentration of iron ions arises in the oxidation of iron in the layer adjacent to the electrode. The formation of iron (II) hydroxide in the electrolysis of the solutions cited is indicated by the fact that stripping curves of the deposit with one maximum of current are also observed after carrying out the electrolysis at -1.0 V, at which reduction of iron (III) ions only to the divalent state and precipitation of the corresponding hydroxide of the element are possible. A direct proportionality is observed between the maximum electrochemical oxidation current of the element and the ion concentration over a wide range of iron (III) ion concentrations.

The determination of iron is strongly disturbed by copper and lead found even in commensurate amounts. The disturbing effect of lead, copper, and other heavy metals is eliminated by introducing several milligrams of sodium sulfide into the solution being analyzed. In the presence of tartrate or citrate ions, heavy metals precipitate as sulfides, but iron (III) ions remain in the solution and are deposited in the form of the metal on the graphite electrode.

Techniques have been proposed to determine iron in ammonium molybdate, citric acid, and tartaric acid with the use of graphite (type I) and saturated calomel electrodes /84/.

Determination of iron in ammonium molybdate. The preparation of the ammonium molybdate sample to be analyzed is roasted 4 hr in an open muffle furnace at 600°C. Then 1 g of the calcined substance is dissolved in 20 ml of a solution containing 85 g of KOH, 1 g of tartaric acid, and 0.1 g of Na_2S per liter. The solution is transferred to the electrolytic cell, oxygen is removed by an inert-gas current, and the electolysis of the stirred solution carried out at a graphite electrode potential of -1.4 V for 15 min. The stirring is stopped and the anodic polarization curve is recorded. Iron oxidizes at from -0.9 to -0.7 V. The iron concentration is determined from a known addition.

In the determination of $1 \cdot 10^{-5}\%$ iron in ammonium molybdate, the coefficient of variation is 10%.

The technique may be used for the analysis of sodium molybdate, sodium tungstate, and ammonium tungstate.

*Determination of iron in citric and tartaric acids.** First, 2 g of the acid sample are dissolved in 15 ml of water, the solution is neutralized to pH 10 to 11 with sodium (potassium) hydroxide, 5—10 mg of crystalline sodium sulfide are added, and the solution volume is brought up to 20 ml with water. The solution is placed in the electrolytic cell, inert gas is passed, and the electrode surface is mechanically cleaned. The electrolysis of the stirred solution is carried out for 15 min at -1.8 V. The stirring is stopped, the electrode potential varied (manually) to -1.2 V, and the anodic polarization curve recorded at from -1.2 to -0.3 V. The maximum anodic current of iron, which is observed at a potential of -0.9 V, is measured.

---

* The technique was proposed by Kh.Z.Brainina and E.M.Roizenblat.

INTERACTION OF METALS DEPOSITED
TOGETHER ON THE ELECTRODE

A mixed metallic deposit forms in the discharge of metal ions on a solid electrode. The features of the stripping of such deposits depend on the nature and interaction of the parent metals. The Me_1—Me_2 systems are divided into three types according to the features of the electrochemical dissolution /85/.

Type I are deposits for which stripping gives evidence of no interaction between the elements. In this case, the anodic polarization curves have two current maxima proportional to the concentration of the corresponding ions. The current maxima are observed on the polarization curve at potentials corresponding to the maximum stripping current of each metal. This type includes deposits containing silver—bismuth, bismuth—lead, lead—cadmium, and copper—bismuth. Despite the relative proximity of the maximum anodic current potentials of lead and cadmium, the lead stripping current is not distorted even for eighty- to hundredfold excess of cadmium. The large difference between the potentials for lead and bismuth stripping allows recording of the bismuth stripping current in the presence of a four-hundredfold excess of lead. The anodic current of bismuth decreases by a total of 20% relative to the value obtained after the electrochemical deposition of the metal from a solution not containing lead. The anodic current of lead remains virtually unchanged when the ratio [Pb (II)] : [Bi (III)] in the solution varies from 1:0 to 1:200. In the silver—bismuth system, interaction of the elements is not observed for the ratios [Ag (I)] : :[Bi (III)]　1:20 and [Bi (III)] : [Ag (I)]　1:25.

This group should also include the copper—bismuth system, in spite of the fact that in most electrolytes the electrochemical dissolution potentials of copper and bismuth are almost identical. Use of a supporting electrolyte solution 1 N in potassium thiocyanate and 1 N in nitric acid allows us to obtain distinct anodic currents for these metals at -0.4 and -0.1 V. The maximum currents for the ratio [Cu (II)] : [Bi (III)] ranging from 1:200 to 20:1 virtually coincide with the corresponding values for the pure metals. However, for higher amounts of copper, a current peak independent of the copper content in the deposit is observed in the potential range for the electrochemical dissolution of bismuth.

The data in the table show that metals forming deposits of this type have different crystalline structures and do not combine in intermetallic compounds.

TABLE. Some properties of metals and two-metal systems

System	Lattice structures of the individual metals /87/	Atomic diameter, $\overset{\circ}{\mathrm{A}}$	Formulas of known intermetallic compounds /88/	Type of electro-chemical dissolution
Ag—Bi	(cubic face-centered)—(rhombohedral)	2.88—3.64	—	I
Bi—Pb	As—Cu	3.64—3.50	Pb_3Bi (presumably)	I
Pb—Cd	Cu—Zn	3.50—3.04		I
Cu—Bi	Cu—As	2.56—3.64		I
Ag—Cu	Cu—Cu	2.88—2.56		II
Cu—Pb	Cu—Cu	2.56—3.50		II
Cu—Co	Cu—Mg ((hexagonal close packing) or Cu	2.56—2.468	—	II
Cu—Cd	Cu—Zn (close to hexagonal close packing, strained Mg structure)	2.56—3.04	Cu_2Cd Cu_4Cd Cu_5Cd_8 $CuCd_3$	III
Sb—Pb	As—Cu	3.22—3.50	(existence of the compound is disputed)	III
Sb—Cd	As—Zn	3.22—3.04	CdSb Cd_3Sb_2	III
Sb—Sn	As—β-Sn (tetragonal)	3.22—3.072	SnSb	III
Ag—Sn	Cu—β-Sn	2.88—3.072	Ag_3Sn	III
Cu—Fe	Cu—α-Fe (or Cu)	2.56—2.56	—	III

Type II are systems in which the interaction appears at commensurate amounts of metals in the deposit such as silver—copper, copper—lead, and copper—cobalt. In this case, the maximum current of the electronegative element decreases as its relative concentration decreases. An additional anodic current peak arises on the polarogram at a potential midway between the potentials of the anodic current peaks of the pure metals (Figure 27). Similar dependences are observed in alternating current polarograms. The additional peak is usually not observed in the successive deposition of the metals.

Tindall and Bruckenstein /86/ used a platinum "ring disk-electrode" to study the electrochemical dissolution of silver and copper. A deposit formed on the disk was anodically stripped and the dependence of the current through the ring on the disk potential, which was varied linearly with time, was recorded. The ring electrode was given a potential corresponding to the limiting diffusion current of the Ag and Cu^{2+} ions. The supporting electrolyte solution was 0.2 M sulfuric acid. The authors showed that an interaction between the metals is not observed in the dissolution of a monolayer deposit. When the amount of deposit exceeds the monolayer, copper dissolves first at from -0.10 to +0.20 V. Then, apparently a solid solution of copper and silver is stripped at from +0.20 to +0.55 V and finally silver, at from +0.55 to +1.0 V.

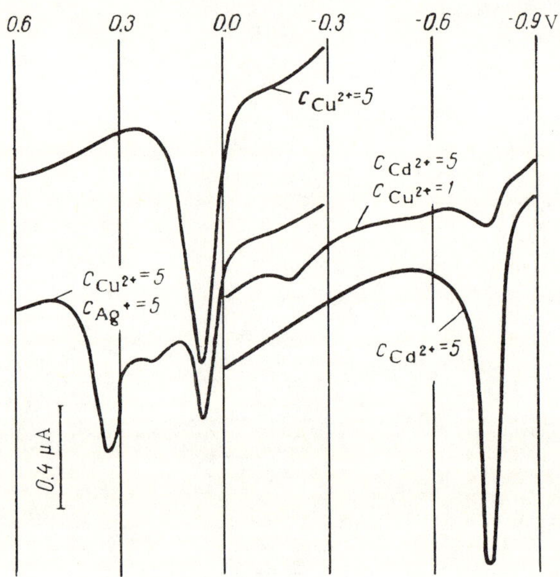

FIGURE 27. Polarization curves for the stripping of deposits from solutions with different metal ion concentrations (numbers on the curves: $\times 10^{-6}$ g-eq/liter).

This system type, as the table shows, includes metals with similar crystal lattices which do not form intermetallic compounds. In this case, solid solutions evidently form in the simultaneous deposition of the metals. An interaction between the metals probably appears more strongly when their atomic radii are similar. For example, the interaction between the elements in the copper—silver system is considerably more pronounced than in the copper—lead system. Oxidation of a solid solution of a metal in a metal having a more positive potential occurs in the intermediate potential range and is recorded in the form of an additional peak on the polarization curve. Formation of solutions of an electropositive element in an electronegative one evidently does not influence the anodic polarization curves, since the solid solution is destroyed when the electronegative component (base) is oxidized.

Type III includes systems in which an interaction between the metals is observed in the presence of very small amounts of the more electropositive component in the deposit such as copper—cadmium, lead—antimony, antimony—cadmium, antimony—tin, silver—tin, and copper—iron. Figure 27 shows the anodic polarization curve (electrochemical dissolution of a deposit containing cadmium and copper) characteristic of this type of system. In these systems, a decrease in the dissolution current of the metal having the more negative potential is observed even when there are small amounts of the metal with the more positive potential in the deposit. The value of the anodic current maximum for cadmium decreases by a factor of two when the ratio $[Cd^{2+}]:[Cu^{2+}]$ equals $1:0.1$ and vanishes when the ratio $[Cd^{2+}]:[Cu^{2+}]$ is less than or equal to $1:0.2$. These ratios remain unchanged when the deposition time is varied and consequently are independent of the absolute amount of copper and cadmium on the electrode within the limits studied. The interaction between the elements in the cadmium—copper system is the strongest of all those studied. This example clearly shows that the decrease or disappearance of the dissolution current of one of the elements in this group is not accompanied by the appearance of any intermediate peaks on the curve or an increase in the anodic current of the second element.

The interaction between cadmium and copper in successive electrochemical deposition appears to a considerably lesser extent than in the simultaneous deposition. The order in which the layers are arranged hardly influences the nature of the observed dependences. Similar phenomena are observed in the Sb—Pb, Sb—Cd, and Sb—Sn systems.

The electrochemical dissolution of a deposit containing iron and copper is distinguished by the fact that one of the peaks on the polarization curve disappears depending on the quantitative relation between the components. The oxidation current of iron remains virtually unchanged only when $[Fe\,(III)]:[Cu\,(II)]$ is greater than or equal to 4.5 and the iron excess represses the copper peak. The copper oxidation current shows conventional behavior when $[Cu\,(II)]:[Fe\,(III)] \geqslant 0.5$. The iron current decreases and vanishes in the case in which $[Cu\,(II)]:[Fe\,(III)] > 5$.

A different situation arises in the electrochemical dissolution of an iron—copper mixture deposited layer by layer. If the inner layer is copper, larger stripping currents for iron are observed than those found after deposition of the element on a clean graphite surface, and the metals do not interact with each other. The copper substrate previously formed on the graphite electrode facilitates crystallization of the iron, thus promoting electrodeposition of iron from dilute solutions. In the electrochemical dissolution of a deposit in which iron forms the inner layer, a decrease in the anodic current of iron is observed due to the screening effect of the deposited copper.

The third type includes systems in which formation of an intermetallic compound is possible. The table shows that according to data in the literature such compounds form in all cases except copper—iron and, possibly, antimony—lead systems. The features of the electrochemical dissolution of deposits of this type are evidently a consequence of the formation of intermetallic compounds in the electrolysis step. The behavior of a deposit containing iron and copper (as well as antimony and lead), is not an exception since, in the electrochemical crystallization of mixed metallic deposits, the formation of solid solutions and intermetallic compounds not appearing on the state diagrams of the corresponding systems is possible /89/.

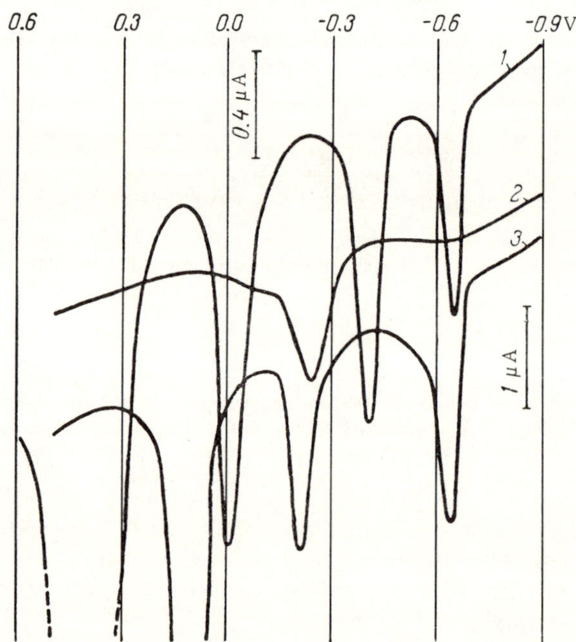

FIGURE 28. Stripping curves for metals deposited at $\varphi_{el} = -0.9$ V and $\tau_1 = 1$ min:

1) Cd, Pb, Cu, and Hg deposited simultaneously from an 0.5 N KNO_3 solution containing $2.5 \cdot 10^{-6}$ g·ion/liter of Cd^{2+}, $2.5 \cdot 10^{-6}$ Pb^{2+}, $7.5 \cdot 10^{-6}$ Cu^{2+}, and $5 \cdot 10^{-5}$ Hg^{2+}; 2) Cd, Pb, and Cu (without Hg) deposited from an 0.5 N KCl solution containing $1 \cdot 10^{-5}$ g·ion/liter of Cd^{2+}, $2.5 \cdot 10^{-6}$ Pb^{2+}, and $1 \cdot 10^{-5}$ Cu^{2+}; 3) Cd and Cu are first deposited and then Hg from an 0.5 N KCl solution containing $1 \cdot 10^{-5}$ g·ion/liter of Cd^{2+}, $1 \cdot 10^{-5}$ Cu, and $5 \cdot 10^{-5}$ Hg^{2+}.

Furthermore, the interaction between metals is most intense in simultaneous electrochemical crystallization which is conducive to formation of a common crystal lattice. In the successive deposition of the metals, this interaction is considerably reduced.

The behavior relating the electrochemical dissolution of mixed metallic deposits, the crystal structure of the metals, and the ability to form intermetallic compounds makes the use of anodic stripping voltammetry of metals a promising method for studying the nature of systems formed during electrolysis. Distortions in the stripping curves of several metals forming type II and type III systems may be eliminated by introducing Hg^{2+} ions.

Figure 28 shows the anodic stripping curves of cadmium, lead, and copper deposited simultaneously with mercury (1). The effect of the interaction of the metals on the electrode disappears upon introduction of $5 \cdot 10^{-6}$ g-ion/liter Hg (II) in the case of type II systems and $5 \cdot 10^{-5}$—$5 \cdot 10^{-4}$ g-ion/liter Hg (II) in type III systems, which amounts to a ten- to hundredfold molar excess of mercury in the solution. After mercury deposition on an electrode on which cadmium and copper has previously been deposited, the usual cadmium anodic peak is observed on the polarization curve (3), whereas without the deposition of mercury the cadmium peak does not appear on the polarogram (2). An analogous technique enables us to detect the stripping currents of tin from a deposit containing silver and tin, and cadmium from a deposit containing antimony and cadmium. Complete elimination of the interaction is possible only by simultaneously depositing mercury. Evidently, the mercury deposited during the electrolysis prevents construction of a common metal crystal lattice and the metal interaction. The electrode created in the electrolysis process is preferable to the mercury film electrode used in amalgam polarography with accumulation /90/ in that the small amount of mercury deposited on the graphite hardly reduces the large working potential range of the graphite electrode. Only the interval from 0.1 to 0.2 V, in which mercury stripping occurs, is excluded from this range. Special preparation of the electrode is also obviated.

Chapter III

STRIPPING VOLTAMMETRY OF VARIABLE-VALENCE IONS

Stripping voltammetry of variable-valence ions consists in recording the polarization curves for the electrochemical dissolution or an electrochemical conversion of a deposit of a chemical compound previously concentrated on the electrode which is formed from the ions being determined.

FORMATION OF INSOLUBLE COMPOUNDS ON THE SURFACE OF AN INDIFFERENT ELECTRODE

Concentrating variable-valence ions on the surface of an electrode is accomplished by electrochemical oxidation or reduction of the ions being determined in the presence of compounds or ions capable of combining with the electrode reaction product to form an insoluble compound. In general, the process may be represented by the following equations /56, 91/:

$$Me^{n+} \longrightarrow Me^{(n \pm m)+} \pm me \qquad (III.1)$$

$$Me^{(n \pm m)+} + (n \pm m)A^{-} \longrightarrow MeA_{(n \pm m)} \downarrow \qquad (III.2)$$

where Me^{n+} and $Me^{(n \pm m)+}$ are different valence states of the element Me and A^{-} is a substance (hydroxyl ions, inorganic or organic substances) which forms an insoluble compound with the $Me^{(n \pm m)+}$ ions.

The upper signs refer to anodic reactions, and equations with the lower signs describe a cathodic process.

The formation of a deposit consists of several steps, namely, transport of Me^{n+} ions from the solution bulk to the electrode surface, transfer of electrons, removal of the ions formed to the solution bulk where these ions combine to form an insoluble compound.   Depending on the relationship between the rates of the chemical reaction (III. 2) and the removal of the ions formed as a result of oxidation or reduction, the insoluble compound forms on the electrode surface or in the solution bulk.   Therefore, a necessary condition for concentrating $Me^{(n\pm m)+}$ ions on the electrode-solution interface is a rather high rate for the chemical reaction binding these ions in an insoluble compound relative to the rate of the removal of the $Me^{(n\pm m)+}$ ions into the solution bulk.

The reagents used and the reactions must satisfy the following requirements.

1. The reagent must form an insoluble compound only with the valence form of the ions being determined which forms as a result of the electrode reaction.   The ions being determined in the solution under study must be fairly stable chemically.

2. The reagent used must not be oxidized or reduced in the potential range for deposition and determination of the element being investigated.

3. The compound arising from conversion of the element being determined must be as insoluble as possible in water, chemically stable in the supporting solution used, and fairly active electrochemically.

4. The conditions for the chemical reaction and the electrochemical conversion of the element being determined into the required valence form must be compatible, i. e., the pH of the medium, the presence of masking agents and auxiliary reagents to keep the element being determined in the initial valence state must not interfere with the necessary electrode reaction in the working potential range used.

5. The chemical reaction for the formation of the insoluble compound must be much faster than the removal of the ions formed in this reaction from the electrode surface.   In order to minimize kinetic difficulties in the course of the chemical reaction, it is obviously wise to use reactions of very low order and select reagents not having groups which sterically hinder the formation of the precipitate.

6. The concentration of the reagent must be adequate for the formation of the deposit on the electrode from solutions with very low concentrations of the ions being determined and for providing the necessary reaction rate.

7. The forces binding the deposit to the electrode surface must be adequate to hold the deposit on this surface.

Under these conditions, the rate of deposition is determined by the magnitude of the electrochemical oxidation or reduction current for the ions deposited which, in turn, depends on the rate of transporting these ions to the electrode surface, the rate of the electrochemical conversion, and the rate of binding the reaction products in the insoluble compound. Since the deposition is carried out at the potential for the limiting diffusion current of the electroactive ions from the stirred solution, a steady state is usually established.

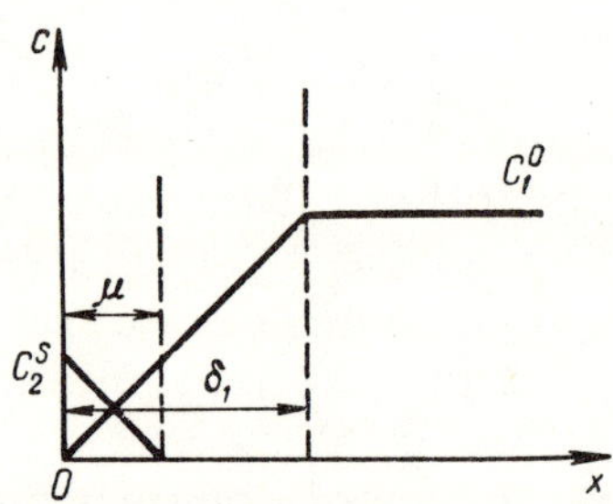

FIGURE 29. Distribution of the concentrations of ions participating in an electrode process.

Figure 29 is a schematic representation of the possible distribution of the concentrations of substances participating in the electrode process after establishment of the steady state /92/. Here, c_1 is the concentration of Me^{n+} ions, which transform into $Me^{(n\pm m)+}$ as a result of the electrochemical oxidation or reduction. The change in the concentration of Me^{n+} ions from c_1^0 (concentration in the solution bulk) to zero near the electrode surface occurs in the diffusion layer of thickness δ, which is determined by the stirring conditions in the solution. The concentration of $Me^{(n\pm m)+}$ ions resulting from the electrode reaction is obviously maximalized and equal to c_2^s near the electrode surface and decreases as the insoluble compound forms in the layer of thickness μ /6, 24—26/. The concentration of these ions in the solution bulk is close to zero. In the case of a fast chemical reaction, the removal of ions by diffusion may be neglected. The flux of the Me^{n+} ions at the electrode surface equals the corresponding flux of the $Me^{(n\pm m)+}$ ions with the opposite sign:

$$D_1 \left.\frac{\partial c_1}{\partial x}\right|_{x=0} = -D_2 \left.\frac{\partial c_2}{\partial x}\right|_{x=0} \qquad \text{(III.3)}$$

where D_1 and D_2 are the diffusion coefficients of Me^{n+} and

$Me^{(n\pm m)+}$ ions, and $\left.\dfrac{\partial c_1}{\partial x}\right|_{x=0}$ and $\left.\dfrac{\partial c_2}{\partial x}\right|_{x=0}$ are the corresponding

concentration gradients near the electrode surface.

In the steady state, the concentration gradient of diffusing ions is a constant and is determined by the diffusion layer thickness and the difference in the concentrations of the respective ions on the boundary of the layer.

The expressions for the concentration gradients of Me^{n+} and $Me^{(n\pm m)+}$ ions may thus be written as

$$\left.\frac{\partial c_1}{\partial x}\right|_{x=0} = \frac{c_1^0}{\delta} \quad \text{and} \quad -\left(\frac{\partial c_2}{\partial x}\right)\bigg|_{x=0} = \frac{c_2^s}{\mu} \tag{III.4}$$

Hence, using relation (III.3) and assuming $D_1 = D_2$, we easily obtain a relation between the concentrations of $Me^{(n\pm m)+}$ ions near the electrode surface and Me^{n+} in the solution bulk:

$$\frac{c_1^0}{\delta} = \frac{c_2^s}{\mu} \tag{III.5}$$

The formation of an insoluble compound deposit is obviously possible if the inequality

$$c_2^s \cdot c_{A^-}^{(n\pm m)} \geqslant K_{sp} \tag{III.6}$$

is fulfilled, where c_{A^-} is the concentration of component A^- which forms an insoluble compound with the electrode reaction product $Me^{n\pm m}$ and K_{sp} is the solubility product.

Since component A^- is introduced into the solution in adequate amounts, the concentration c_{A^-} may be assumed constant throughout the entire solution and in the region adjacent to the electrode. Substituting relation (III.5) into (III.6), we obtain a tentative expression for evaluating the minimum concentration $c_{1\,min}^0$ of Me^{n+} ions at which those ions may be concentrated on the electrode as an insoluble compound of ions of the same element with a different valence:

$$c_{1\,min}^0 \geqslant \frac{K_{sp}}{c_{A^-}^{(n\pm m)}} \cdot \frac{\delta}{\mu} \tag{III.7}$$

When $n \pm m = 1$

$$c_{1\,min}^0 \geqslant \frac{K_{sp}}{c_{A^-}} \cdot \frac{\delta}{\mu}$$ (III. 8)

Hence, the minimum concentration c_1^0 (which amounts to the sensitivity of the method) is determined by the solubility product of the precipitate formed, the concentration of A^-, and the ratio of δ and μ. Since μ cannot be larger than δ (otherwise the precipitate would form in the solution bulk rather than on the electrode surface), c_1^0 should be somewhat larger than the solubility product indicates.

When μ is very small and δ large (corresponding to a large concentration of A^- and weak stirring of the solution), the minimum concentration determined is much higher than might be expected from the value of the solubility product of the compound formed on the electrode. Thus, the relation between the amount of the compound formed on the electrode and the electroactive ion concentration may be expressed as

$$Q = \begin{cases} 0 \text{ when } c_1^0 < c_{1\,min}^0 \\ f(c_1^0) \text{ when } c_1^0 \geqslant c_{1\,min}^0 \end{cases}$$ (III. 9)

The function $Q = f(c_1^0, \tau_1)$ over a wide range of values for c and τ_1 may be represented schematically as in Figure 30. The segment $0A$ corresponds to values of c_1^0 less than or equal to $c_{1\,min}^0$. Here, the precipitate does not form. When c_1^0 is greater than $c_{1\,min}^0$, steady-state conditions for the formation of the precipitate exist as long as the solution is not depleted or a very thick film of the compound does not form on the electrode. The first limitation arises in all types of electrodeposition and has been often discussed /93/. The extent of solution depletion is proportional to the ratio of the electrode surface to the solution volume, and under normal experimental conditions (for example, a solution volume of 20 ml, an electrode area of 0.02 cm^2, and an electrolysis time of $\leqslant$ 30 min) depletion of the solution may be neglected. The second limitation arises due to the deposition of a solid compound on the electrode surface. Depending on whether this compound is a conductor, semiconductor, or insulator, the kinetics of the process will vary with time. If the compound is a conductor, an electrochemical reaction may occur on the compound-solution interface, and the rate of the process

should be determined by the factors already described.   There
is experimental evidence that the amount of compound which
forms on the electrode in the deposition process in this case is
a linear function of the product of the electroactive ion concentration and the electrolysis time /94/.   This dependence is
represented graphically by the line **AD** (in the general case, the
segments **AB** and **BD** may have different slopes).

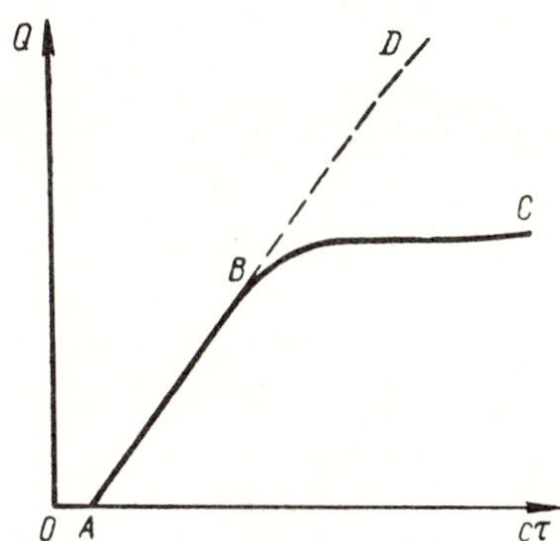

FIGURE 30.   The amount of compound proportional to Q
formed on the electrode as a function of the value of $c\tau_1$.

In the case of a deposit with high electrical resistance, an
additional slowing down of the process occurs owing to the slowed
passage of ions participating in the electrode reaction through
the deposit layer.   When the rate of this new step in the process
becomes commensurate with (or less than) the rate of diffusion,
the current decreases.   This state, which is somewhat similar to
making the electrode passive, obviously sets in faster the
higher the resistance of the substance formed, the lower its poros-
ity, and the greater the bulk concentration of the ions participating in the electrode reaction.   The kinetics for the formation
of compound with high resistance is described by the curve **ABC**.
As long as the quantity of the substance on the electrode is small,
we may expect the quantities Q and $C_1^0\tau_1$ to be directly proportional.   When the amount of the compound increases as a result
of increased resistance, a drop in the current and a decrease in
the rate of deposition should be observed.   In the limiting case,
the rate should vanish.   The length of interval **AB** obviously depends on the nature, structure, and electrical properties of the
precipitate.   No precipitate is formed within the **BC** segment.

Hence, the relation between the amount of substance formed
on the electrode and the concentration of the electroactive ions
and the electrolysis time may be expressed by the equations:

$$Q = nFSD \; \frac{c_1^0 - c_{1\,min}^0}{\delta} \; \tau_1 \qquad \text{(III. 10)}$$

or, if $c_1^0 \gg c_{1\,min}^0$

$$Q = nFSD \; \frac{c_1^0 \tau_1}{\delta} \qquad \text{(III. 11)}$$

These equations are valid for the case in which the deposit of
the insoluble compound which forms on the electrode does not
have a resistance greater than the diffusion resistance for transporting the reacting ions to the electrode surface.   Equations
(III. 10) and (III. 11) describe the relationship between the quantity
of compound which forms on the electrode, the concentration of
electroactive ions, and the electrodeposition time only in the
simplest case not complicated by any specific effect of the
electrolyte on the film.

STRIPPING OF COMPOUNDS FROM THE SURFACE
OF AN INDIFFERENT ELECTRODE

The stripping of compounds localized on the surface of an indifferent electrode obviously depends on the electrochemical
properties of the film.   Reduction (or oxidation) of compounds
with fairly high electrical conductivity does not require the
electroactive substance to be brought to the electrode surface
beforehand.   The kinetics of such a process is determined by
the relationship between the rates of electron transfer and removal of the reaction products to the solution bulk.   Since this
kinetics is mathematically analogous to the case of stripping a
metal from the surface of an indifferent electrode described
above, it will not be given special consideration here.
The dissolution of deposits of compounds with low electrical
conductivity apparently occurs through a prior ionization step
accompanied by the transport of electroactive ions to the electrode surface.   Dissolution preceded by a chemical step is
described by the equations /91/:

$$D \frac{\partial^2 c_1}{\partial x^2} + k_1 c_2 - k_2 c_1 = 0 \qquad \text{(III. 12)}$$

$$\frac{\partial c_2}{\partial t} = k_2 c_1 - k_1 c_2 \tag{III.13}$$

with the following initial and boundary conditions:

$$c_2(x, 0) = c_2^0 ; \qquad c_1(x,0) = \frac{c_2^0}{K'}$$

$$\left.\frac{\partial c_1}{\partial x}\right|_{x=0} = c_1(0,t)k \exp(Bvt); \qquad \left.\frac{\partial c_1}{\partial x}\right|_{x=l} = 0 \tag{III.14}$$

where c_1 and c_2 are the concentrations of substances $Me^{n\pm m}$ and $MeAn\pm m$; k_1, the rate constant of the forward chemical reaction $MeA_{(n\pm m)} \underset{k_2'}{\overset{k_1}{\rightleftarrows}} Me^{(n\pm m)+} + (n \pm m)A^-$; k_2, the effective rate constant of the reverse reaction $k_2 = k_2' c_A^{n\pm m}$, and $K' = \dfrac{k_2}{k_1} = \dfrac{c_2^0}{c_1^0}$.

For a cathodic process:

$$B = -\frac{anF}{RT}; \qquad k = \frac{k_S}{D_1} \exp\left[-\frac{anF}{RT}(\varphi_1 - \varphi^0)\right]$$

For an anodic process:

$$B = \frac{\beta nF}{RT}; \qquad k = \frac{k_S}{D_2} \exp\left[\frac{\beta nF}{RT}(\varphi_1 - \varphi^0)\right]$$

The derivative of the concentration with respect to time has been omitted in equation (III.12) since only chemical reactions faster than the diffusion process are considered ($k_1 t, \ k_2 t \gg 1$). Equation (III.13) does not contain a term describing the diffusion of the solid $MeA_{n\pm m}$ since its diffusion may be neglected. The boundary conditions reflect the fact that the $Me^{(n\pm m)+}$ ions participating in the electrode process do not enter the film from the outside and their concentration gradient at the electrode surface is proportional to the stripping current.

The expressions for the maximum stripping current and the potential of the polarization curve maximum have the form:*

$$i_{max} = \frac{nFDS\gamma Qa_\infty \sqrt{v}}{Kf_{1\mu}\sqrt{\pi k_1 \varphi_m}\left(1 + \dfrac{1}{2B\varphi_m}\right)} \tag{III.15}$$

$$i_{max} = \frac{(nFDS)^2 a_\infty \gamma\sqrt{v}}{Kf_{1\mu}\sqrt{\pi k_1 \varphi_m}\ \delta}(c_1^0 - c_{1\,min}^0)\tau_1 \tag{III.16}$$

* For solution of this problem, see Appendix IV.

$$i_{max} = \frac{(nFDS)^2 a_\infty \gamma \sqrt{v}}{K f_1 \mu \delta \sqrt{\pi k_1 \varphi_m}}\, c_1^0 \tau_1 \qquad (III.\ 17)$$

$$\varphi_{max} = \frac{1}{2B}\ln \varphi_m - \frac{1}{B}\ln \frac{k\mu\sqrt{\pi k_1}}{2B} + \frac{1}{2B}\ln v \qquad (III.\ 18)$$

Hence, the maximum current for stripping a compound from the surface of an indifferent electrode depends on the chemical reaction equilibrium and rate constants, the hydrodynamic conditions, and the electrode potential scanning rate in the stripping step. The maximum current is directly proportional to the amount of the compound localized on the electrode surface, the deposition time, and the concentration of the electroactive ions, when the latter is much higher than the minimum concentration at which formation of the precipitate is possible. When the ion concentration is close to the minimum concentration required for precipitation, a linear dependence between the maximum current and the concentration of the electroactive ions is observed.

APPLICATION OF STRIPPING VOLTAMMETRY
IN THE ANALYSIS OF VARIABLE-VALENCE IONS

The direct dependence of the maximum stripping current on the concentration of the electroactive ions allows application of stripping voltammetry in analysis. Increasing the electrolysis time and the electrode potential scanning rate in the stripping cycle enables us to increase the sensitivity of the determination, when the sensitivity is not limited by the solubility product of the compound which forms on the electrode (see equations (III. 8), (III. 16) and (III. 17)).

The possibility of forming various compounds as a result of cathodic and anodic reactions allows analysis of single-component systems without the preliminary separation and isolation of the element being determined and enables us to concentrate and determine elements when analysis by stripping voltammetry of metals is difficult or impossible.

The precipitating reagents used in stripping voltammetry of variable-valence ions include hydroxyl ions for precipitation of hydroxides /56, 95—101/, other inorganic ions for precipitation of salts /92/, and various organic compounds /102—107/. The use of organic precipitating agents is specially recommended

when these agents are selective and the compounds formed
are insoluble.

The selectivity of stripping voltammetry of variable-valence
ions depends on the selectivity of the precipitating agent used
and the individual electrochemical properties of the element
being determined. All ions which do not form insoluble compounds in solution and are electroactive in the potential range
applied do not interfere with the determination. The determina-
tion of cerium employing the reaction $Ce\,(III) + 4OH^- \rightleftharpoons Ce(OH)_4 + e$
is not disturbed by rare earth elements which are not electro-
active from $+1.0$ to $+0.3$ V relative to the saturated calomel
electrode where the oxidation-reduction reactions of cerium
ions occur /98/. If the concentrating step is accomplished by
a reduction reaction, the determination is not disturbed by trace
elements present in the solution in their lowest valence state.
If the concentrating step is the result of an anodic process, the
determination is not disturbed by more electropositive elements,
since the net reaction may be carried out at a potential insufficient for their oxidation. The determination of iodide ions em-
ploying the reaction $2I^- + Cl^- + R \rightleftharpoons R[I_2Cl] + 2e$ (R^+ is the rhodamine S cation) is virtually not disturbed by any element in-
cluding any amounts of chlorides and bromides /106/.

If the electroactive ions in the initial valence state form
sufficiently insoluble compounds with the components of the
solution, a new solid phase may arise on the electrode in the
process which reverses the concentrating step. For example,
the reduction of iron (III) hydroxide results in the formation of
a $Fe(OH)_2$ precipitate which may be oxidized during anodic polarization of the electrode. In such cases, the cyclic polarograms have a cathodic and an anodic current maximum /108/.
Similar phenomena are observed in the nickel—dimethylglyoxime
system /107/. The experimental data indicate that the maximum
current for the electrochemical conversion and the amount of
precipitate on the electrode are directly proportional.

CONCENTRATION AND DETERMINATION OF
ELEMENTS AS INSOLUBLE HYDROXIDES
AND INORGANIC SALTS

Stripping voltammetry of variable-valence ions may be used
to concentrate and determine elements forming hydroxides in

different valence states.   Let the element Me be characterized
by the two valence states Me^{n+} and $Me^{(n+m)+}$, and the $Me^{(n+m)+}$
ions form the less soluble hydroxide.   If the hydrogen ion concentration is such that the Me^{n+} ions do not precipitate out,
and the $Me^{(n+m)+}$ ions undergo hydrolysis, the following oxidation
process may occur on the electrode with formation of a hydroxide
precipitate:

$$Me^{n+} \longrightarrow Me^{(n+m)+} + me$$
$$Me^{(n+m)+} + (n+m)OH^- \longrightarrow Me(OH)_{n+m} \downarrow$$

Thus, the concentrating step will occur on the electrode surface.

A linear varying potential on the cathodic side causes stripping of the hydroxide:

$$Me(OH)_{n+m} \longrightarrow Me^{n+m} + (n+m)OH^-$$
$$Me^{(n+m)+} + me \longrightarrow Me^{n+}$$

These equations describe an anodic concentrating step and
cathodic stripping of the precipitate.

Similarly, the reverse processes may be represented in those
rare cases in which the hydroxide of the lower oxidation state of
the element being determined is less soluble.

The application of this method in the determination of cerium,
thallium, iron, manganese, lead, rhenium, and chromium is considered.

Cerium

Cerium is concentrated on the electrode as a result of the
oxidation of Ce^{3+} ions and hydrolysis of the electrode reaction
products.   Here a precipitate of the hydroxide or a basic salt
of tetravalent cerium forms on the electrode-solution interface.
The anodic polarization curves for cerium in 0.1 N acetate buffer
are well defined.   The limiting current of the Ce^{3+} ions is attained at 0.8 V, and thus the element may be concentrated as the
hydroxide at a more positive electrode potential.   The optimal
electrolysis potential is +1.0 V.   At lower potentials, accumulation of cerium hydroxide on the electrode does not occur, while

concentrating at more positive potentials is accompanied by distortion in the cathodic polarization curve probably as the result of the accumulation of oxygen on the electrode.

The polarization curves for the reduction of cerium (IV) hydroxide, obtained at varying electrolysis times in solutions containing varying amounts of cerium nitrate, have a characteristic shape with a well-defined maximum.   The value for this maximum is directly proportional to the electrodeposition time and the Ce^{3+} ion concentration in the $1 \cdot 10^{-6}$—$1 \cdot 10^{-4}$ g-ion/liter range.   In prolonged electrolysis (15—20 min) of solutions containing more than $5 \cdot 10^{-5}$ g-ion/liter of Ce^{3+}, the electrode becomes passive and the current ceases to depend on the cerium ion concentration and the electrolysis time.

A similar dependence between the cerium ion concentration and the maximum stripping current for cerium (IV) hydroxide is also observed in the presence of large amounts of other rare earth elements and allows determination of trace amounts of cerium in rare earth compounds.

This method for cerium determination is very sensitive and selective.   In commensurate amounts, only manganese interferes.

The mean square error in the determination of $5 \mu g$ Ce (III) in 30 ml of solution is $0.2 \mu g$, and the coefficient of variation equals 4%.

Techniques have been proposed for the determination of cerium in rare earth elements /98/ and zinc employing graphite (type I) and saturated calomel electrodes.

The determination of cerium is based on the reaction

$$Ce^{3+} + 4OH^- \rightleftarrows Ce(OH)_4 \downarrow + e$$

Determination of cerium in rare earth elements /98/.   First, 2 g of the metal sample are dissolved in nitric or hydrochloric acid, several drops of a hydrogen peroxide solution are added, the solution is boiled until the reducing agent decomposes completely, and the volume of the solution is brought up to 20 ml with water.   The solution is transferred to the electrolytic cell and oxygen removed by bubbling nitrogen.   Anhydrous sodium acetate is added until the pH is between 4 and 5 and the electrolysis of the stirred solution is carried out for 10 min at + 1.0 V. Cerium (IV) hydroxide forms on the electrode.   The stirring is stopped and the cathodic polarization curve for the stripping of

cerium (IV) hydroxide is recorded at from $+1.0$ to 0.0 V.   The
cathodic current is measured and the cerium ion concentration is
found by using a calibration graph.

In the determination of $1 \cdot 10^{-4}\%$ cerium, the coefficient of
variation does not exceed 10%.

*Determination of cerium in zinc.** First, an 0.1—1 g zinc
sample is dissolved in 10 to 20 ml hydrochloric acid (1:1), the
solution evaporated to dryness, the residue dissolved in 15—20 ml
0.1 M acetate buffer with gentle heating without boiling, the solution is cooled and transferred to the electrolytic cell.   Oxygen
is removed by bubbling inert gas for 10—15 min.   The electrolysis of the stirred solution is carried out for 10—15 min at
$+1.0$ V.   Cerium (IV) hydroxide forms on the electrode.   The
stirring is stopped and the cathodic polarization curve for the
stripping of cerium (IV) hydroxide is recorded.   The maximum
cathodic current, which is observed at a potential of approximately $+0.3$ V, is measured.   The cerium ion concentration is found
from a calibration graph.

In the determination of $1 \cdot 10^{-4}\%$ cerium in zinc the coefficient
of variation does not exceed 20%.

Thallium

Thallium is concentrated on the electrode as a result of the
oxidation of Tl^+ ions and hydrolysis of the electrode reaction
product.   A precipitate forms consisting of $Tl_2O_3 \cdot nH_2O$ and a
small amount of anions /109, 110/.   The electrochemical oxidation of Tl^+ ions on the graphite electrode occurs at more
positive potentials the lower the pH.   The half-wave potentials
of the anodic polarization curves are given in Table III. 1.   The
anodic waves of thallium are well defined.   The direct dependence of the limiting current on the Tl^+ ion concentration allows
determination of relatively high concentrations (greater than
$1 \cdot 10^{-3}$ g-ion/liter).   The determination of smaller amounts is
possible after electrochemically concentrating the element on
the electrode as thallium (III) hydroxide.

---

* The technique was proposed by E.Ya.Neiman and G.P.Dolgopolova. See "Korroziya i
  elektrokhimiya" (Corrosion and Electrochemistry), Proceedings of the "Gidrotsvetmeto-
  brabotka" Institute, report No.31, Metallurgiya, 1970, p.125.

TABLE III.1. Conditions for the oxidation of Tl^+ and the reduction of thallium (III) hydroxide [$Tl + 3OH^- \rightleftarrows Tl(OH)_3 + 2$]

Solution composition	pH	$\varphi_{1/2}$	φ_{el}	φ_{max}	c_{min} g-ion/ /liter
0.35 M $(NH_4)_2SO_4 + H_2SO_4 + Tl^+$	4	+1.25	+1.5	+0.5	$1 \cdot 10^{-5}$
0.35 M $(NH_4)_2SO_4 + Tl^+$	5	+1.15	+1.4	0.35	$2 \cdot 10^{-6}$
0.35 M $(NH_4)_2SO_4 + NH_4OH + Tl^+$	7	+1.0	+1.2	+0.3	$5 \cdot 10^{-7}$
0.35 M $(NH_4)_2SO_4 + NH_4OH + Tl^+$	8	+0.8	+1.1	+0.2	$5 \cdot 10^{-7}$
0.35 M $(NH_4)_2SO_4 + NH_4OH + Tl^+$	9	+0.7	+0.9	+0.15	$3 \cdot 10^{-6}$
0.35 M $(NH_4)_2SO_4 + NH_4OH + Tl^+$	10	+0.55	+0.8	+0.1	$9 \cdot 10^{-5}$

Polarization curves for the electrochemical reduction of thallium (III) hydroxide have a sharply defined current maximum, which is shifted in the negative direction when the pH of the solution is increased. The most symmetric and sharply defined polarization curves for the electrochemical reduction of thallium (III) are found for neutral solutions. These conditions are optimal for electrode-position of the element. The maximum cathodic current for the stripping of thallium (III) hydroxide is directly proportional to the amount of hydroxide on the electrode, the electrodeposition time, and the thallium (I) ion concentration in the 10^{-5}—10^{-4} g-ion/liter concentration range. At lower thallium (I) ion concentrations, there is a linear relationship between the stripping current of thallium (III) hydroxide and the Tl^+ concentration (Figure 31). The minimum concentration that can be determined is $5 \cdot 10^{-7}$ g-ion/liter (see Table III. 1).

The determination of thallium in the presence of large amounts of copper is extremely interesting. In classical and amalgam polarography, an electrode process involving copper ions precedes or occurs simultaneously with the reduction of thallium ions, and when the amounts of copper are increased, the copper masks the oxidation or reduction current for thallium. Since copper ions are usually present in the solution in the divalent state and do not undergo further oxidation, a fairly large quantity of this element does not interfere with the determination of thallium by the method considered in this section. When the electrode potential scanning rate is slow, a sharper resolution of the currents due to the electrode processes in the reduction of thallium (III) hydroxide and Cu^{2+} ions is observed but the net signal usually decreases.   Therefore, an average voltage scanning rate of 0.015 V/sec is preferable in the determination of thallium in the presence of large amounts of copper.   Polarization curves for the stripping of thallium (III) hydroxide obtained at this electrode potential scanning rate in the presence of increasing amounts of copper are shown in Figure 32.   As seen from this figure, the polarization curve for thallium is not distorted in the region of the maximum even when the [Tl] : [Cu] ratio is $1 : 1.25 \cdot 10^5$ (0.5 M $CuSO_4$).   The increase in the maximum stripping current for thallium (III) (baseline at the end of the left branch of the curve), which is observed when the Tl^+ ion concentration is increased, indicates that the determination of thallium is possible in the presence of large amounts of copper, as well as of copper compounds.   The sensitivity of the determinations in the latter case for an electrolysis time of 5 min in $7 \cdot 10^{-4}\%$ (1.5 g $CuSO_4$ in 20 ml of solution).   Sensitivity may be increased to $5 \cdot 10^{-5}\%$ if the electrode is transferred from the copper-containing solution to a pure supporting electrolyte solution after accumulation of and subsequent recording of the cathodic polarization curve for the stripping of thallium (III) hydroxide.   In this case, the amount of copper salt sample may be increased and a smaller stripping current for thallium (III) hydroxide may be recorded.   This method of determining thallium is fairly sensitive and accurate. The coefficient of variation does not exceed 15%.

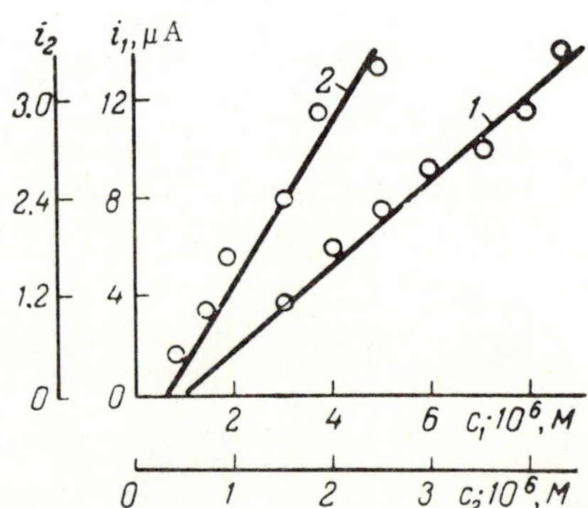

FIGURE 31. Dependence of the maximum stripping current for thallium (III) hydroxide on the Tl^+ concentration in an ammonia solution 0.5 M in $(NH)_4SO_4$ under the conditions:

1) v = 0.125 V/sec, τ_1 = 2 min, pH = 9; 2) v = 0.018 V/sec, τ_1 = 7 min, pH = 8.3.

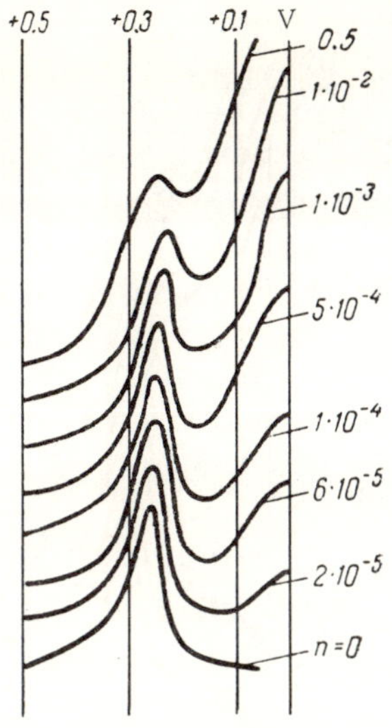

FIGURE 32. Polarization curves for the stripping ($v = 0.015$ V/sec) of thallium (III) hydroxide deposited at $\varphi_{el} = 1.0$ V, $\tau_1 = 5$ min from ammonia solutions (pH = 9) 0.5 M in $(NH_4)_2SO_4$, $4 \cdot 10^{-6}$ M in Tl_2SO_4, and n M in $CuSO_4$ (numbers on the curves).

Use of the anodic reaction $Tl^+ + 3OH^- \longrightarrow Tl(OH)_3 + 2e$ allows determination of thallium in indium. For this purpose, the metallic indium sample is dissolved in sulfuric acid, a solution of $(NH_4)_2SO_4$ and NH_4OH is added, the solution is heated, the precipitate filtered off, and thallium is determined in the filtrate. The polarization curves in this case are the same as in Figure 32. Thallium (I) is not trapped in considerable amounts in the indium hydroxide precipitate.

The determination of thallium in the presence of lead which may be oxidized and accumulated on the electrode as a hydroxide is a more difficult problem. However, thallium may be determined in a mixture of elements which precipitate in an ammonia-ammonium buffer. Coprecipitation of lead (II) hydroxide occurs in this buffer and virtually undistorted polarization curves are obtained when the filtrate is analyzed.

Compounds which are soluble in an ammonia-ammonium buffer solution and are not electroactive at from + 1.0 to 0 V relative to the saturated calomel electrode do not interfere with the determination of thallium.

A technique for the determination of thallium in cadmium sulfate employing graphite (type I) and saturated calomel electrodes has been proposed.

The determination is based on the reaction

$$Tl^+ + 3OH^- \rightleftharpoons Tl(OH)_3 \downarrow + 2e$$

Determination of thallium in cadmium sulfate /100/. First, a 6 g cadmium sulfate sample is dissolved in 30 ml 2.5 M ammonium hydroxide, the pH is brought down to 8.5, and the solution is transferred to the electrolytic cell. The electrolysis of the stirred solution is carried out for 3 min at a graphite electrode potential of +0.95 V. Thallium (III) hydroxide forms on the electrode. Then the cathodic polarization curve for the stripping of thallium (III) hydroxide is recorded at from +0.95 to 0 V. The maximum cathodic current is measured. The concentration of thallium (I) in the solution is found by using a calibration graph.

In the determination of $1 \cdot 10^{-4}\%$ thallium in a cadmium salt, the coefficient of variation is 10%.

The determination is not disturbed by large amounts of copper, indium, zinc, nickel, magnesium, and alkali metals.

Iron

The fact that the divalent and trivalent ions of iron form compounds with different solubilities in an alkaline medium is exploited to concentrate and determine iron. There are two possibilities in the concentrating step, namely, concentrating iron found in solution in the divalent state as trivalent iron hydroxide or vice versa /108/.

When the concentrating step from a sodium tetraborate solution involves deposition of $Fe(OH)_3$ on a hanging mercury drop electrode, the experiment is carried out in the following manner to prevent oxidation of iron (II) in the alkaline medium. A buffer solution is placed in the electrolytic cell and oxygen is removed by passing an inert gas. The complete removal of the oxygen and the stirring of the solution are monitored by the cathodic current at -0.4 V. When the current falls to $0.1 \mu A$, freshly prepared acidified Mohr's salt solution is introduced. Deposition of iron hydroxide is carried out at -0.05 V from the stirred solution. Then the cathodic polarization curve for reduction of the precipitate is recorded.

The optimal pH value for the buffered solution used is selected by considering that, on the one hand, increasing hydroxyl ion concentration shifts the equilibrium in the required direction and the rate of formation of $Fe(OH)_3$ on the electrode increases, but on the other hand, the $Fe(OH)_2$ precipitation in the solution bulk

becomes possible.   The optimal value found for the pH is 8.
Iron (II) solutions not containing oxygen are sufficiently stable.

The polarization curves for the electrolytic reduction of iron
(III) hydroxide are well defined.   The maximum cathodic current
(φ_{max} = -0.5 V) is directly proportional to the concentration of the
iron (II) ions in the $1 \cdot 10^{-7} - 5 \cdot 10^{-5}$ M range.

A nonlinear dependence is observed between the maximum
current for the electrochemical reduction of iron (III) hydroxide
and the formation time.   Such a dependence often has been noted
in amalgam vector polarography with accumulation /111/ and
has been explained by the effect of adsorption of surface-active
substances on the mercury electrode.   When a graphite electrode is used under similar conditions, no appreciable repression
of the process has been detected in vector polarography.   Obviously, using a graphite electrode is also preferable to using a
hanging mercury drop electrode in the stripping voltammetry of
variable-valence ions.

This method for iron determination is fairly selective.   The
determination of iron is not disturbed by hundredfold excesses
of copper, lead, nickel, cobalt, and thallium, since these elements
are not concentrated on the electrode under these conditions.
The polarization curves remain unchanged when noble metals,
cadmium, zinc, and more electronegative elements are present
in the sample.   The minimum concentration of iron which may
be determined is $1 \cdot 10^{-7}$ M.

Sample preparation is as follows.   After reduction in a reducing cell, an acidic solution of the sample is collected in the
electrolytic cell containing a weighed amount of boric acid.
Oxygen is removed from the solution and NaOH is added.   The
solution must have a pH between 8 and 8.3.   The subsequent
procedure is described above.

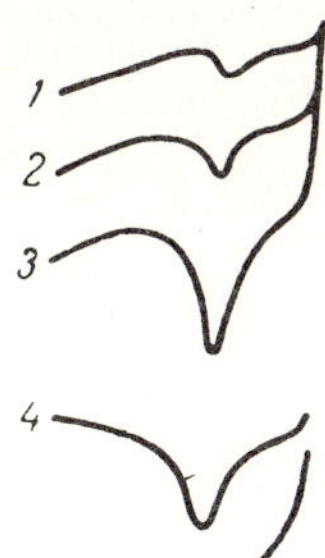

FIGURE 33.   Polarization curves for the stripping
of iron (II) hydroxide deposited at φ_{el} = -1.0 V,
τ_1 = 2 min (curves 1—4) and 5 min (curve 5) and
different Fe^{3+} concentrations:

1, 4, 5) $2 \cdot 10^{-6}$; 2) $4 \cdot 10^{-6}$; 3) $6 \cdot 10^{-6}$ g-ion/liter.

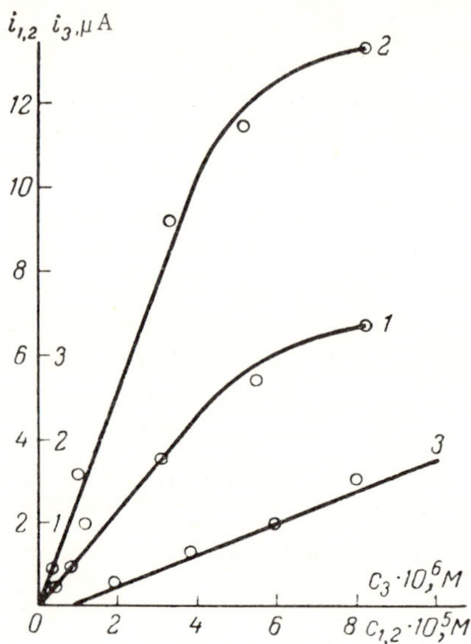

FIGURE 34. Dependence of the maximum anodic current for the stripping of iron (II) hydroxide on the Fe^{3+} concentration in an alkaline (pH = 10) 0.5% $C_3H_4OH(COO)_3^{3-}$ at $\tau_1 = 2$ min (curves 1, 3) and 5 min (curve 2).

The variant involving trivalent iron ions and deposition of insoluble iron (II) hydroxide is convenient. The solution being investigated is more stable than in the previous case. Iron is concentrated from alkaline 0.5% citrate buffer (pH = 10) at -1.0 V. The half-wave potential of the corresponding cathodic polarogram is about -0.85 V. Stripping of the deposit formed here is described by a polarization curve with one well-defined anodic current maximum at -0.7 V (Figure 33). The maximum current is linearly dependent on the iron (II) ion concentration in the $2 \cdot 10^{-6}$ to 10^{-5} M range and directly proportional in the $1 \cdot 10^{-5}$ to $5 \cdot 10^{-5}$ M range (Figure 34). More concentrated solutions of iron (III) are unstable under these conditions.

The determination of iron employing the reaction $Fe^2 \rightleftarrows Fe^{3+} + e$ with the deposition of an insoluble hydroxide is far more sensitive and selective than the methods of classical polarography.

Manganese

Manganese is concentrated on a graphite /96/ or platinum /95, 112/ electrode as a result of the oxidation of Mn^{2+} ions employing the reaction /113—116/ $Mn^{2+} + 4OH^- \rightleftharpoons Mn(OH)_4 \downarrow + 2e$. The limiting anodic current is proportional to the concentration of Mn^{2+} ions and is practically independent of the pH of the medium /96/. The half-wave potential is shifted in the positive direction when the pH decreases. The oxidation wave of manganese in 0.1 N nitric acid is masked by the anodic current of the supporting electrolyte.

The optimal potential values for precipitating manganese (IV) hydroxide on a graphite electrode /96/ from solutions with varying pH values are given in Table III. 2. Reduction of manganese (IV) hydroxide in acid medium is described by almost symmetrical polarization curves characterized by a sharp rise and a fall of the current virtually to the background line. The descending branch of the polarization curves does not reach the baseline in the case of more alkaline solutions. In a repeated cathodic cycle, the residual current is significantly higher than on a clean electrode. Repeated electrochemical and mechanical cleaning is necessary to restore the electrode to its original state. Larger amounts of deposits on the electrode lead to increased distortion of the curves. In such cases, the curve is not symmetrical even for an acidic solution.

TABLE III.2. Conditions for oxidation of Mn^{2+} and reduction of manganese (IV) hydroxide

Solution composition	pH	$\varphi_{1/2}$	φ_{el}	φ_{max}
0.1 M HNO$_3$ + Mn^{2+}		—	1.3	0.9
2 M (NH$_4$)$_2$SO$_4$ + H$_2$SO$_4$ + Mn^{2+}	3	0.9	1.1	0.7
2 M (NH$_4$)$_2$SO$_4$ + H$_2$SO$_4$ + Mn^{2+}	4	0.8	1.0	0.65
2 M(NH$_4$)$_2$SO$_4$ + H$_2$SO$_4$ + Mn^{2+}	5	0.75	0.9	0.25
2 M (NH$_4$)$_2$SO$_4$ + NH$_4$OH + Mn^{2+}	7	0.5	0.7	0.15
2 M (NH$_4$)$_2$SO$_4$ + NH$_4$OH + Mn^{2+}	8	0.25	0.5	0.10
2 M (NH$_4$)$_2$SO$_4$ + NH$_4$OH + Mn^{2+}	9	0.15	0.4	0.05

The complicated nature of the dependence of the maximum electrochemical reduction current on the concentration of Mn^{2+} ions in the solution is apparently due to the difference in the anodic and cathodic processes. The mechanism of the cathodic process depends on the acidity of the medium. The reduction of manganese (IV) hydroxide is not always accompanied by dissolution of the precipitate /96, 112, 115—121/. The cause of this behavior may lie in the reduction of the precipitate to insoluble hydroxide MnOOH as an intermediate which is converted into the soluble divalent state of manganese only in a subsequent electrochemical or disproportionation reaction. In neutral solutions, the reduction reaction product is a completely or partially insoluble compound which causes a distortion of the polarization curves and a bend in the calibration graph.

When weakly acidic ammonium sulfate solutions are used as the supporting electrolyte, symmetrically shaped electroreduction polarization curves can be obtained with a sharply defined current maximum directly proportional to the manganese ion concentration. Thus, the reaction $Mn^{2+} \rightleftharpoons Mn^{4+} + 2e$ may be utilized for concentrating and determining traces of manganese. The sensitivity of the Mn^{2+} determination is $5 \cdot 10^{-7}$ g-ion/liter.

The determination of manganese is not disturbed by alkali metals, alkaline earths, zinc, or cadmium. No interference is observed for a thousandfold excess of copper, a hundredfold excess of thallium, or a commensurate amount of lead. The concentration of iron and aluminum should not exceed $1 \cdot 10^{-5}$ g-ion/liter. Tartrates and versene interfere at concentrations greater than 0.01 M.

Techniques for manganese determination in cadmium sulfate /96/ and alkali metal hydroxides /112/ have been developed.

The determination is based on the reaction

$$Mn^{2+} + 4OH^- = Mn(OH)_4 \downarrow + 2e$$

Determination of manganese in cadmium sulfate /96/.
Graphite* (type I) and saturated calomel electrodes are employed.

First, an 8 g cadmium sulfate sample, weighed to an accuracy of 0.01 g, is dissolved in 15 ml water, the pH is brought to 5 with ammonium hydroxide, and the solution diluted with water to 20 ml. The sample solution is transferred to the electrolytic cell, and oxygen is removed by bubbling nitrogen for 15 min. The electrolysis of the stirred solution is carried out at +0.9 V. Manganese is concentrated on the electrode as manganese (IV) hydroxide. Then the stirring is stopped and the cathodic polarization curve for the electrochemical reduction of manganese

---

* The electrode is mechanically cleaned between measurements.

(IV) hydroxide is recorded at from $+0.9$ to 0 V. The maximum cathodic current is measured. The Mn^{2+} concentration is determined by using a calibration graph.

In the determination of $2.5 \cdot 10^{-5}\%$ manganese in cadmium sulfate, the coefficient of variation is 10%.

Determination of manganese in alkali metal hydroxides /112/. A platinum or a graphite (type I or III) electrode and saturated calomel electrodes are used.

First, a 4 g hydroxide sample is dissolved in 100 ml 1.5 N ammonium chloride. The solution is transferred to an electrolytic cell and oxygen removed by bubbling inert gas. The electrolysis of the stirred solution is carried out for 15 min at $+0.6$ V. Manganese is concentrated on the electrode as manganese (IV) hydroxide. Then the stirring is stopped, the cathodic polarization curve for the electrochemical reduction of manganese hydroxide is recorded, and the maximum cathodic current is measured. The manganese concentration is found from a calibration graph.

Lead

Pb^{2+} ions are oxidized on platinum /122/ and graphite /123/ electrodes.

The anodic waves are well defined. The limiting current is proportional to the lead ion concentration. The half-wave potential is shifted from $+0.5$ to $+1.4$ V in the transition from alkaline to acidic solutions. Conditions for the oxidation of Pb^{2+} and the reduction of lead (IV) hydroxide are:

pH of the solution ...	13	11	9	7	5	3	1
$\varphi_{1/2}$	0.5	0.7	0.9	1.1	1.3	1.4	1.4
φ_{el}	0.7	0.9	1.0	1.25	1.5	1.6	1.65
φ_{max}	0	0.15	0.3	0.45	0.6	0.7	0.8

In acidic solutions, the oxidation potentials of the lead ions and the supporting electrolyte are close. However, the segment for the anodic current of lead is fairly well defined in differential polarization curves. The electrochemical oxidation of lead ions is accompanied by the formation of insoluble lead (IV) hydroxide on the electrode.

When the electrode potential is shifted in the negative direction, lead (IV) dioxide (hydroxide) is reduced. The corresponding polarization curves have a current maximum. The potential of the maximum is dependent on the pH of the medium, and the maximum cathodic current is proportional to the amount of precipitate on the electrode, the concentration of lead (II) ions in the $1 \cdot 10^{-7} - 1 \cdot 10^{-6}$ g-ion/liter range, and the oxidation time.

The determination of lead is not disturbed by substances electrochemically inactive in the positive potential region. Nickel, copper, and antimony in fifty to hundredfold excesses do not interfere.

In the determination of $1 \cdot 10^{-7}$ g-ion/liter Pb^{2+}, the coefficient of variation is 5—6%.

Rhenium

Rhenium may be concentrated electrochemically as a film /124/ which according to Demkin and Sinyakova is ReO_2. The concentrating step is carried out on a mercury drop at -0.9 V. The maximum stripping current is observed at -0.7 V. The best supporting electrolyte solution is 4—5 M phosphoric acid.

The maximum of the anodic current derivative with respect to time is directly proportional to the rhenium concentration in the $5 \cdot 10^{-9} - 1 \cdot 10^{-7}$ M range for accumulation times of 3 to 5 min. At higher rhenium ion concentrations, the peak bifurcates and the dependence becomes complicated.

The determination of rhenium is not disturbed by copper and lead in five-hundred-thousandfold excesses, molybdenum in ten-thousandfold excesses, or tungsten in less than a hundredfold excess. NO_3^- ions interfere with the determination of rhenium.

Chromium

Concentrating chromium as a hydroxide
Chromium is concentrated on the graphite electrode in the reduction of CrO_4^{2-} ions and hydrolysis of the electrode reaction product. The cathodic polarograms have the usual shape.

Examples of integral and differential anodic curves are presented in Figure 35. Table III.3 gives the half-wave potential values for the reduction of chromium (IV) ions on a graphite electrode in different solutions and the potentials of the anodic polarization curve maxima for the electrochemical oxidation of the hydroxide (or basic salt) of chromium (III).

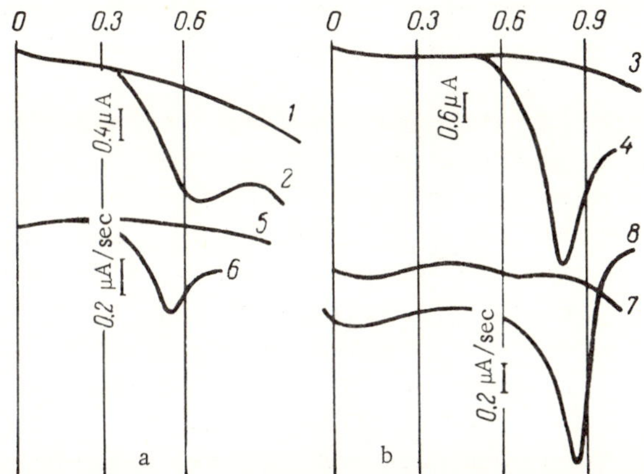

FIGURE 35. Integral (1—4) and differential (5—8) anodic polarization curves for the stripping ($v = 0.011$ V/sec) of the hydroxide or basic salt of chromium (III) deposited when $\tau_1 = 1$ min (curves 2, 4, 6, and 8) and $\tau_1 = 0$ (curves 1, 3, 5, 7):

a) from a solution 0.05 M in $Na_2B_4O_7$ and $1 \cdot 10^{-4}$ M in Na_2CrO_4 at $\varphi_{el} = -1.0$ V;
b) from acetate buffer (pH = 4.7), $[CrO_4^{-2}] = 1 \cdot 10^{-4}$ g-ion/liter at $\varphi_{el} = -0.5$ V.

TABLE III.3. Conditions for the reduction of CrO_4^{2-} ions and the oxidation of $Cr(OH)_3$
$[CrO_4^{2-} + 4H_2O + 3e \rightleftarrows Cr(OH)_3 + 5OH^-]$

Solution composition	pH	$\varphi_{1/2}$	φ_{el}	φ_{max}	c_{min} g-ion/ /liter
0.1 M NaOH + CrO_4^{2-}		—0.8	—1.0	+0.17	$1 \cdot 10^{-4}$
0.05 M $Na_2B_4O_7$ + CrO_4^{2-}	9.2	—0.7	—1.0	+0.6	$2 \cdot 10^{-7}$
$C_6H_8O_7$ + NaOH + CrO_4^{2-}	5.2	—0.35	—0.5	+0.7	$1 \cdot 10^{-6}$
KH_2PO_4 + NaOH + CrO_4^{2-}	8.0	—0.35	—0.7	+0.7	$1 \cdot 10^{-7}$
0.4 M NH_4Cl + 0.1 M NH_4OH + CrO_4^{2-}		—0.45	—0.7	+0.6	$4 \cdot 10^{-8}$
0.1 M CH_3COOH + 0.1 M CH_3COONa + CrO_4^{2-}		—0.1	—0.5	+0.9	$2 \cdot 10^{-8}$
0.1 M H_2SO_4 + Cr_2O_7		+0.15	—0.3	+1.15	$1 \cdot 10^{-5}$

The position of the polarization curves relative to the potential axis is dependent on the pH of the solution. The electrochemical reduction of CrO_4^{2-} proceeds most easily (at more positive potentials) in acidic solution (pH = 2.5). When the pH is increased, the waves are shifted toward the negative potentials. The limiting current in all cases is determined by the CrO_4^{2-} concentration in the solution. This dependence is nonlinear apparently due to the rapid formation of a fairly thick layer of the hydroxide or basic salt of chromium (III) in solutions containing a relatively large amount of CrO_4^{2-}.

The prior concentrating of chromium on the electrode at potentials adequate for the reduction of CrO_4^{2-} ions is accompanied by the appearance on the polarization curve of a characteristic peak which is more pronounced the lower the pH. When alkaline solutions are used, the peak is poorly defined as shown in the left part of the figure, and the current for the oxidation of Cr^{3+} is partially masked by the background oxidation current. The lower part of the figure shows the differential curves for the electrochemical oxidation of the precipitate. In both cases, the maximum of the anodic current derivatives with respect to time is well defined and increases as the electrode potential is shifted in the negative direction in the deposition cycle. The optimal potentials (φ_{el}) for concentrating chromium from different solutions are given in Table III.3. These potentials coincide in each case with the region of the corresponding limiting current for the electrochemical reduction and are shifted in the negative direction as the hydroxyl ion concentration is increased. It should be noted that the half-wave potentials of the cathodic polarograms and of the anodic polarization curve maxima differ by 1—1.4 V as evidenced by the fact that the electrode process is essentially irreversible.

The almost directly proportional relationship observed between the maximum of the anodic current derivative with respect to time and the chromate (or dichromate) ion concentration allows determination of the chromium concentration. The maximum sensitivity of the determination is attained when ammonia-ammonium buffer is employed. In this medium, chromium may be concentrated from a solution containing $4 \cdot 10^{-8}$ g-ion/liter chromium. The minimum sensitivity of the determination is obtained when solutions 1 N in NaOH or $K_2SO_4 + H_2SO_4$ (pH = 2.5) are used as the supporting electrolyte. Concentrating and determining chromium from weakly acidic and weakly alkaline solutions are possible in the 10^{-6}–10^{-5} g-ion/liter CrO_4^{2-} concentration range. The higher sensitivity of the chromium determination from alkaline buffer solutions is due to the high concentration of hydroxyl ions which are the precipitating agents. The decrease in the sensitivity of the determination in a solution of 1 N NaOH is probably due to the amphoteric properties of chromium (III) and the related increase in the solubility of chromium (III) compounds in alkaline solutions.

The well-defined pH dependence of the oxidation potential of Cr^{3+} ions and the reduction potential of CrO_4^{2-} ions enables us to select the optimal solution composition for analysis of composite mixtures.

The coefficient of variance in the determination of $0.05\,\mu g/ml$ chromium equals 7%. Chromium may be determined by this method on a background of alkali metal salts in the presence of large amounts of some elements. The directly proportional relationship between the maximum of the anodic current derivative with respect to time and the Cr (IV) ion concentrations is observed, for example, in the presence of large amounts of zinc sulfate and ammonium tungstate (or molybdate). Commensurate amounts of copper, antimony, and bismuth, as well as alkaline earths in amounts not precipitating CrO_4^{2-} do not interfere with the determination. Metal ions which undergo hydrolysis in the buffer solutions used do not interfere as long as the precipitate does not trap chromate ions. For example, Al^{3+} may be present in large amounts.

The determination of chromium is disturbed by lead, silver, mercury, and tin. However, by complexing the interfering element, we may determine chromium not only in the presence of large excesses of the element but also in compounds of the element (for example, in lead molybdate). Techniques have been proposed to determine chromium in ammonium molybdate, lead molybdate, and zinc sulfate employing graphite (type I) and saturated calomel electrodes.

The determination of chromium is based on the reaction

$$CrO_4^{2-} + 4H_2O + 3e^- \rightleftharpoons Cr(OH)_3 \downarrow + 5OH^-$$

*Determination of chromium in ammonium molybdate.** The sample is roasted for 4 hr in an open muffle furnace at 600°C.

* The technique was proposed by T.A.Krapivkina and Kh.Z.Brainina.

Then 1 g of the calcined sample is dissolved in 10 ml 10% alkali, two or three drops of hydrogen peroxide are added, and the solution is stirred and boiled for 10 min. The solution is neutralized with nitric acid to pH 8, the volume of the solution is brought up to 25 ml with water, the solution is transferred to the electrolytic cell, and oxygen is removed by bubbling inert gas. The electrolysis of the stirred solution is carried out for 15 min at -1 V. Chromium is concentrated on the electrode as $Cr(OH)_3$. The voltage applied to the cell is decreased to 0 V, and the differential anodic polarization curve for the electrochemical oxidation of chromium hydroxide is recorded at from 0 to +0.7 V. The maximum of the anodic current derivative with respect to time is measured. The concentration of chromium is found from a calibration graph.

In the determination of $1 \cdot 10^{-4}\%$ chromium, the coefficient of variation is 10%.

Sodium molybdate, sodium tungstate, and ammonium tungstate may be analyzed by this method.

Determination of chromium in lead molybdate. First, a 1 g lead molybdate sample is triturated with 5—6 ml 10% sodium hydroxide, 20 ml of this solution is added, and the mixture is boiled until the sample dissolves completely. Three or four drops of hydrogen peroxide are added and the solution is stirred and boiled for another ten minutes. Then, 1.5 g versene and 3 ml concentrated nitric acid are introduced, and the solution is brought to pH=8 with 1 N nitric acid and diluted to 25 ml with water. The solution is transferred to the electrolytic cell and oxygen is removed by bubbling inert gas. The electrolysis of the stirred solution is carried out for 30 min at -1.0 V. The stirrer is stopped and the graphite electrode transferred to an electrolytic cell filled with 0.05 M sodium tetraborate. The electrode with the deposit is kept out of solution when the current is turned off during the transfer. The differential anodic polarization curve for the electrochemical oxidation of chromium hydroxide is recorded at from 0 to +1.0 V. The maximum of the anodic current derivative with respect to time is measured. The concentration of chromium is found from a calibration graph.

In the determination of $5 \cdot 10^{-5}\%$ chromium, the coefficient of variation is 20%.

*Determination of chromium in zinc sulfate.** First a 1 g zinc sulfate sample is dissolved in 20 ml 1 M ammonium hydroxide,

---

* The technique was proposed by Kh.Z.Brainina and T.Z.Krapivkina.

several drops of hydrogen peroxide are added, the solution
boiled for 10 min until the hydrogen peroxide decomposes completely, and the volume is brought up to 20 ml with a solution
1 M in ammonium hydroxide and 0.1 M in ammonium chloride.
The solution is transferred to the electrolytic cell and oxygen
is removed by bubbling inert gas. The electrolysis of the
stirred solution is carried out for 5 min at -1.2 V. The stirring
is stopped, the potential shifted to zero, the solution allowed to
settle, and the differential anodic curve for the electrochemical
oxidation of chromium hydroxide is recorded at from 0 to +0.8 V.
The maximum of the current derivative with respect to time is
measured. The chromium concentration is found from a calibration graph.

   In the determination of $1 \cdot 10^{-4}\%$ chromium in zinc sulfate, the
coefficient of variation is 10%.

   Concentrating and determining chromium
   by precipitation as insoluble inorganic salts
   The concentrating electrode reaction results in the deposition of a neutral or basic salt instead of a hydroxide on the
electrode surface. An example of such a concentrating step is
the formation of barium chromate and lead chromate.

   Chromium (III) in low concentrations does not form insoluble
compounds in solutions of barium hydroxide,* barium acetate,
or lead acetate while chromium (VI) precipitates in the form of
the corresponding chromate. This property of the element is
exploited in the concentrating step according to the reactions
/92/:

$$Cr^{3+} + 8OH^- \longrightarrow CrO_4^{2-} + 4H_2O + 3e$$
$$CrO_4^{2-} = Me^{2+} \longrightarrow MeCrO_4 \downarrow$$

where Me^{2+} is Ba^{2+} or Pb^{2+}

   The half-wave potentials for the oxidation of Cr^{3+} ions in
solutions of barium hydroxide and lead acetate are 0.2 and 0.8 V
(Table III. 4). The electrolysis of solutions containing chromium
(III) ions and those of barium or lead at potentials more positive
than the values indicated is accompanied by the formation of the
corresponding metal chromate precipitates on the electrode.
The optimal concentrating potentials correspond to the region of
the limiting current for oxidation of the element (Table III. 4).

* A saturated solution of $Ba(OH)_2$ prepared in a supporting electrolyte freed of CO_2 in a
  nitrogen atmosphere previously passed through NaOH solution is used.

TABLE III.4. Conditions for oxidation of Cr^{3+} and reduction of barium chromate and lead chromate

Solution composition	$\varphi_{1/2}$	φ_{el}	φ_{max}	c_{min} g-ion/ /liter
Saturated solution of $Ba(OH)_2 + Cr^{3+}$	+0.2	+0.3	—0.6	$6 \cdot 10^{-8}$
0.02 M $Pb(CH_3COO)_2 + Cr^{3+} +$ $+ 0.1$ M CH_3COONa	+0.8	+1.0	—0.05	$1 \cdot 10^{-4}$
0.09 M $Ba(NO_3)_2 +$ $+ 0.1$ M $NaCH_3COO + Cr^{3+}$	+0.8	+1.6	—0.6	$1 \cdot 10^{-5}$

The polarization curves for the reduction of barium and lead chromate in the first two cases cited in Table III. 4 are well defined, and in the latter somewhat distorted by the reduction current of adsorbed oxygen. The effect of oxygen is eliminated when the electrode with the deposit is transferred to 0.1 N sodium hydroxide solution after concentrating the chromium and subsequently recording the cathodic curve. In this case, the reduction currents of barium chromate and oxygen are observed in different potential ranges. The maximum reduction current of the compound is a linear function of the Cr^{3+} ion concentration in the $1 \cdot 10^{-5}$—$1 \cdot 10^{-4}$ g-ion/liter range and is directly proportional to the concentration in the $1 \cdot 10^{-4}$—$6 \cdot 10^{-4}$ g-ion/liter range for deposition times up to 5 min. Then, the electrode becomes passive, the amount of deposit in further electrolysis does not increase, and the cathodic current ceases to depend on the Cr^{3+} ion concentration in the solution.

This method can determine traces of chromium in the nitrates of barium, strontium, calcium, the rare earth elements, potassium, sodium, magnesium, and ammonium nitrate. The determination is not disturbed by considerable amounts of copper, cadmium, zinc, and lead. In the determination of $1 \cdot 10^{-5}$ g-ion/liter Cr^{3+}, the coefficient of variation is 14%.

A technique has been developed for the determination of chromium in potassium chloride and sodium chloride employing graphite (type I) and saturated calomel electrodes.

The determination is based on the reaction

$$Cr^{3+} + 8OH^- + Ba^{2+} \rightleftharpoons BaCrO_4 \downarrow + 4H_2O + 3e$$

*Determination of chromium in potassium chloride and sodium chloride.** First, 4 g of the salt sample are dissolved in 20 ml saturated barium hydroxide solution.** The solution is transferred to the electrolytic cell and oxygen is removed by bubbling inert gas. The electrolysis of the stirred solution is carried out for 5 min at +0.3 V. Chromium is concentrated on the electrode as barium chromate. Then the stirring is stopped, the solution allowed to settle, and the cathodic polarization curve for the electrochemical reduction of barium chromate is recorded at from 0 to -1.0 V. The maximum cathodic current, which is observed at about -0.6 V, is measured. The chromium concentration in the sample is found from a calibration graph.

In the determination of $2.5 \cdot 10^{-4}\%$ chromium, the coefficient of variation is 10%.

CONCENTRATING AND DETERMINING ELEMENTS PRECIPITATED AS INSOLUBLE COMPOUNDS WITH ORGANIC REAGENTS

A condition for electrochemically concentrating an element employing an organic reagent is the selectivity of this reagent to the valence state of the ion which forms in the electrode reaction and its indifference to sample ions in the initial valence state. Thus, crystal violet and rhodamine S may be used for the electrodeposition of antimony from a chloride solution since they precipitate $[SbCl_6]^-$ which arises in the anodic oxidation of trivelent antimony ions and do not draw Sb^{3+} into the precipitate. To concentrate the element, the sample solution containing the organic reagent is electrolyzed at an electrode potential sufficient for the oxidation or reduction (conversion to the state which reacts with the reagent) of the ions being determined /102/. The electrode process is accompanied by the formation of an insoluble compound. If the reagent precipitates the element being determined in the higher valence state, the following reactions occur:

$$Me^{n+} \longrightarrow Me^{(n+m)+} + me$$
$$Me^{(n+m)+} + (n+m)RH \longrightarrow MeR_{n+m}\downarrow + (n+m)H^+$$

---

* The technique was proposed by T.A.Krapivkina.

** Barium hydroxide is dissolved by boiling in water. The flask is joined to a trap containing NaOH solution and cooled without access to carbon dioxide. All further operations are carried out in a nitrogen atmosphere cleansed by a sodium hydroxide solution.

If the reagent precipitates the element in the lower valence state, the following reactions occur:

$$Me^{(n+m)+} + me \longrightarrow Me^{n+}$$

$$Me^{n+} + nRH \longrightarrow MeR_n \downarrow + nH^+$$

where Me^{n+} and $Me^{(n+m)+}$ are different valence states of the element Me, RH the organic reagent, and MeR_n or MeR_{n+m} the precipitate which forms on the electrode.

When the electrode potential is shifted in the negative direction in the first case or in the positive direction in the second, the electrode deposit undergoes an electrochemical conversion. The maximum current observed is a measure of the concentration of the ions being determined in the solution.

Important advantages of electrochemically concentrating elements with organic reagents may be high electivity and sensitivity. The former factor depends on the selectivity of the reagent used and the individual electrochemical properties of the element being determined. High sensitivity results when a quantity of the insoluble compound deposited on the electrode of the order of $0.001-0.002\,\mu g$ is adequate to give good results.

The precipitation of complex inorganic anions by organic cations is often employed. Basic dyes in acid medium are used as reagents: methylene violet, brilliant green, crystal violet, rhodamines, and organic hydroxides or salts yielding a fairly heavy cation upon dissociation /105, 106, 125, 126/.

To concentrate and determine elements which exist in a cationic state after electrochemical oxidation or reduction, we use organic acids, diethyldithiophosphates, diethyldithiocarbaminates, etc., i. e., organic compounds forming insoluble ionic associations containing the element being determined.

There are specific reagents for a number of elements with characteristic functional groups which form saltlike or coordination compounds. Typical reagents forming coordination compounds with a number of inorganic ions are 8-hydroxyquinoline, dimethylglyoxime, dithizone, and 1-nitroso-2-naphthol.

The compounds formed should be electrochemically fairly active as well as insoluble and chemically stable, while the reagent used should obviously be stable to the electrochemical oxidation and reduction reactions over a wide range of potentials.

Use of a graphite indicator electrode with a wide working potential range is especially convenient since many oxidation and reduction reactions accompanied by the formation of insoluble compounds with organic reagents occur at positive potentials.

In the following examples of concentrating and determining trace amounts of antimony, iodine, cobalt, and nickel, possible uses of different types of organic reagents in stripping voltammetry of variable-valence ions will be shown.

Antimony

Antimony is concentrated from a chloride solution containing rhodamine S and antimony (III). The solution is electrolyzed with a graphite anode at + 0.8 V relative to a saturated calomel electrode. The antimony ions oxidized under these conditions are concentrated on the electrode as the chloroantimonate of rhodamine S. The subsequent cathodic polarization of the electrode with a linearly varying electrode potential is accompanied by the stripping of the deposit /105/. The cathodic polarization curve has a characteristic shape with a current maximum at + 0.3 V. The shape of the curve and the current maximum depend on the electrolysis potential, the electrolysis time, and the concentrations of chloride ions, rhodamine S, and antimony (III). The maximum cathodic current is directly proportional to the concentrating time when the antimony concentration is low. When the electrolysis time of solutions containing more than $2 \cdot 10^{-6}$ g-ion/liter Sb^{3+} is increased (greater than 10 min), the function $i_{max} = f(\tau_1)$ becomes nonlinear, probably because the electrode becomes passive.

A potential of + 0.8 V is optimal for concentrating antimony. At lower potentials, the oxidation rate of antimony (III) is low, and at more positive potentials, the electrolysis is accompanied by distortion of the cathodic polarization curve, probably due to adsorption of chlorine evolved.

The minimum chloride ion concentration at which concentrating antimony is possible is 0.2 g-ion/liter. The fact that in more dilute solutions of potassium chloride antimony is hardly concentrated at all is apparently due to the difficulty of forming anionic antimony complexes* that react with rhodamine S.

---

* Chloride complexes of antimony may be partially hydrolyzed when the solution acidity is inadequate, however the charge and interaction mechanism of these complexes with cationic rhodamine S (R^+) remain unchanged. Therefore, we shall assume that the $[SbCl_6]^-$ ion formally participates in the reactions.

Increasing the potassium chloride concentration above 0.5 M hardly affects the maximum cathodic current.

The concentration of rhodamine S cations must be sufficient to precipitate the $[SbCl_6]^-$ ions formed. This condition is fulfilled by solutions with rhodamine S concentrations greater than 0.02 g/liter acidified with 0.5 M sulfuric acid. At lower reagent concentrations, the polarization curves are distorted or disappear entirely while larger amounts of precipitating agent do not affect the nature of the polarization curves. The maximum current remains constant.

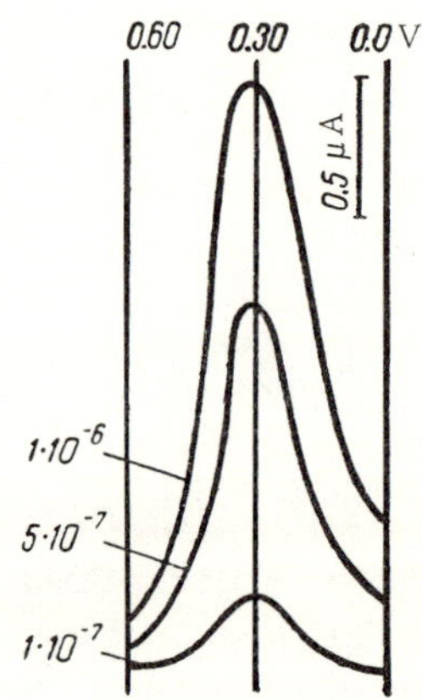

FIGURE 36. Polarization curves for the dissolution of antimony deposited at $\varphi_{el} = 0.8$ V and $\tau_1 = 10$ min from solutions 1.5 M in KCl, $4 \cdot 10^{-4}$ M in rhodamine S, 0.5 M in H_2SO_4, and containing different amounts of Sb (III) (numbers on the curves: g-ion/liter).

The cathodic polarization curves for different antimony concentrations in the solution are illustrated in Figure 36. The dependence of the maximum stripping current on the Sb^{3+} ion concentration is shown in Figure 37. The maximum stripping current is directly proportional to the Sb^{3+} ion concentration.

This method of determining antimony is very sensitive ($5 \cdot 10^{-8}$ g-ion/liter) and very selective. High selectivity is obtained because the electrode reaction used occurs at positive potentials (only a few elements are electroactive at these potentials) and the precipitating agent is quite selective.

Elements fall into several groups according to their effect on the determination of antimony.

1. Elements electrochemically inactive at positive potentials which do not form anionic complexes, and thus do not interact

with rhodamine S and do not interfere with the determination in any amounts.   This group includes alkali metals, alkaline earths, aluminum, rare earth elements, manganese, and thallium (I).

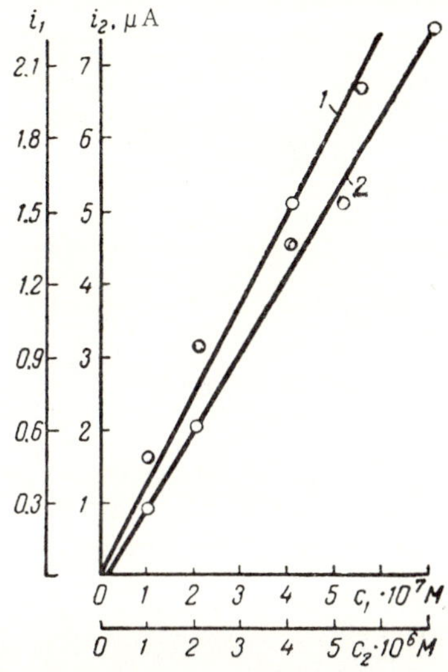

FIGURE 37.   Dependence of the maximum stripping current of rhodamine S chloroantimonate on the antimony ion concentration (for conditions, see Figure 36).

2. Elements electrochemically inactive at positive potentials, which do form anionic complexes capable of reacting with rhodamine S do not interfere when present in quantities inadequate to precipitate or trap microgram amounts of antimony upon precipitation (for example, a six-hundrefold excess of bismuth or a thousandfold excess of copper).   This group includes cadmium, zinc, indium, silver, gallium, mercury, titanium, lead, nickel, and gold (III).

3. Elements electroactive at positive potentials which do not react with rhodamine S.   The oxidation or reduction currents of large amounts of such elements may mask the stripping current of antimony.   In this case, it is wise to carry out the concentrating of antimony on the electrode in the solution being analyzed and the electrochemical dissolution in another solution not containing interfering impurities.

4. Elements electroactive at positive potentials which form anionic complexes precipitated by rhodamine S.   Iodine, for

example, interferes with the determination of antimony even in
commensurate amounts. Nevertheless, as will be shown below,
the determination of antimony is possible.

Techniques have been proposed for the determination of antimony in potassium chloride containing an iodide impurity, in
bronze, and in brass employing graphite (type I) and saturated
calomel electrodes.

The determination is based on the reaction

$$[SbCl_6]^- + R^+ \rightleftharpoons R[SbCl_6] \downarrow + 2e$$

Determination of antimony in potassium chloride. First, a
3 g potassium chloride sample is dissolved in 10 ml 0.6 N hydrochloric acid and 4 ml silver nitrate solution containing 0.01 mg-ion
Ag per ml are added. The precipitate is filtered off (if the
product being analyzed does not contain iodine, this step may be
omitted). Then, 200 mg hydrazine is added to the filtrate which
is gently boiled for 10 min. Then, the solution is cooled, water
is added to 13 ml, and 2 ml 0.15% rhodamine S is added. The
solution is transferred to the electrolytic cell and oxygen is removed by bubbling inert gas. A potential of +0.8 V is established and the cathodic polarization curve is recorded at from
0.8—0 V. The maximum current (i_1) is measured for potentials
ranging from +0.5 to +0.35 V. Electrolysis of the stirred solution is carried out for 30 min at +0.8 V and antimony is concentrated on the electrode as a rhodamine S compound. The stirrer
is turned off and the polarization curve for stripping the deposit
is recorded at from 0.8 to 0 V. The maximum cathodic current
i_2, which is observed at between 0.4 and 0.3 V, is measured.
The antimony concentration is found from the maximum stripping current ($i_2 - i_1$) by using a calibration graph or from a
known addition.* In the analysis of solutions with concentrations
greater than $1 \cdot 10^{-7}$ M Sb^{3+}, the value of i_1 may be neglected.

In the determination of $3 \cdot 10^{-6}$% antimony in potassium chloride, the coefficient of variation is 15%.

This technique may be used to analyze alkali metals, alkaline
earths, and ammonium compounds.

*Determination of antimony in bronze and brass.*** First, an
0.5 g alloy sample is dissolved in 20 ml nitric acid (1:1) and

---

* A standard solution of antimony with an Sb concentration of 0.3 μg/ml is prepared by
   dilution of a 1 N hydrochloric acid solution of Sb_2O_3 in distilled hydrochloric acid.
** The technique was proposed by E. Ya. Neiman.

boiled for 2 to 3 min to remove nitrogen oxides. Then 10 ml hydrochloric acid (1:1) is added and the solution is evaporated to the moist crystal state. The evaporation with hydrochloric acid is carried out five times. The residue is dissolved in 1 N hydrochloric acid, 10^{-3} M in rhodamine S. The solution is transferred to the electrolytic cell and oxygen is removed by bubbling inert gas. Simultaneously, 10 ml hydrochloric acid is poured into a second electrolytic cell and oxygen is also removed by bubbling inert gas.

The electrolysis of the stirred solution is carried out for 10 min at $+0.8$ V and antimony is concentrated on the electrode as a rhodamine S compound. Then, the electrode is carefully immersed in a beaker with doubly distilled water, washed, and transferred to the second electrolytic cell. The cathodic polarization curve for the stripping of rhodamine S chloroantimonate is recorded at once at from 0.8 to 0 V. The maximum cathodic current, which is observed at a potential of about $+0.3$ V, is measured.

The antimony concentration is found using a known addition (the addition is introduced to the first electrolytic cell).

In the determination of $1 \cdot 10^{-4}\%$ antimony, the coefficient of variation is 14%.

Iodine

The oxidation of I^- ions accompanied by the chemical interaction of iodine, chloride ions, and rhodamine S cations is employed /106/ to concentrate iodine on the electrode surface. Polarograms for the oxidation of I^- ions on the graphite electrode in a solution 0.2 M in hydrochloric acid and $5 \cdot 10^{-5}$ M in rhodamine S have the usual form with the half-wave potential about $+0.5$ V. The limiting current is observed at potentials more positive than $+0.7$ V.

Elementary iodine formed in the oxidation is quite soluble and is not concentrated on the electrode (Figure 38). The cathodic polarization curve recorded after electrolysis of the solution at an electrode potential adequate for the oxidation of iodide ions does not have current peaks. Concentrating the element is only accomplished in the presence of rhodamine S and Cl^- ions. Iodine forms the following precipitate:

$$2I^- \longrightarrow I_2 + 2e$$
$$I_2 + Cl^- + R^+ \longrightarrow R[I_2Cl]\downarrow$$

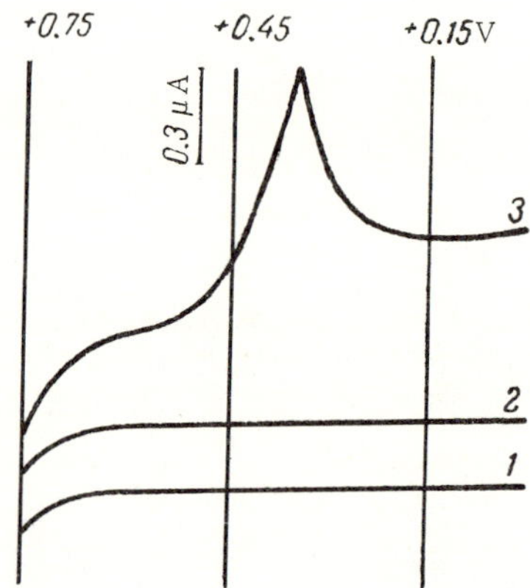

FIGURE 38. Cathodic polarization curves obtained after electrolysis on a graphite electrode at $\varphi_{el} = 0.8$ V and $\tau_1 = 10$ min under the conditions:

1) solution 0.1 M in H_2SO_4, $5 \cdot 10^{-5}$ M in rhodamine S, and $5 \cdot 10^{-7}$ M in KI;
2) solution 0.1 M in H_2SO_4, $5 \cdot 10^{-7}$ M in KI, and 0.1 M in KCl; 3) solution 0.1 M in H_2SO_4, $5 \cdot 10^{-5}$ M in rhodamine S, 0.1 M in KCl, and $5 \cdot 10^{-7}$ M in KI.

A cathodic current peak is observed on the cathodic polarization curve recorded after electrolysis of a solution containing iodide ions, chloride ions, and rhodamine S at an electrode potential sufficient to oxidize iodide ions. The potential of this polarization curve maximum is about + 0.35 V. The precipitate dissolves during the recording of the cathodic polarogram as evidenced by the absence of current peaks after repeated recording.

The shape of the polarization curves and the maximum stripping current depend on the concentration of rhodamine S. The optimal precipitating agent concentration is $5 \cdot 10^{-5}$ M (0.0025%). The oxidation is carried out in a slightly acidic medium which promotes dissociation of rhodamine S and the formation of the organic cations needed for the concentrating step. A large excess of acid in the solution is undesirable because of the chemical oxidation of iodide ions by atmospheric oxygen in acid medium /128/.

Iodine may be concentrated only with an adequate potassium chloride concentration. The chloride ion concentration must be higher than the iodide ion concentration. The best conditions for the concentrating step are created in 0.1 N potassium chloride. When the potassium chloride content is higher, the maximum stripping current remains unchanged but the shape of the cathodic curve is distorted. The potential of the polarization curve maximum is shifted in the negative direction when the chloride ion concentration is increased. The maximum current for the electrochemical dissolution of the compound is directly proportional to the iodide ion concentration in the solution and the deposition time as long as the value of $c\tau_I$ is less than or equal to $1 \cdot 10^{-5}$.

The fact that the calibration graph arises from the coordinate origin indicates the high sensitivity of the method. In fact, it is possible to measure the stripping current of a precipitate after depositing iodine for 15 min from a solution with $c_I = 6 \cdot 10^{-9}$ g-ion/liter.

Recalling relation (III. 8) and that $\delta/\mu > 1$ (implying that fast chemical reaction leads to formation of a precipitate on the electrode surface), we may evaluate the upper limit of the solubility product of the compound which forms on the electrode. In this case, c^0 is less than or equal to $6 \cdot 10^{-9}$ M when $c_A = 5 \cdot 10^{-5}$ M, and thus, K_{sp} is less than $3 \cdot 10^{-13}$. Actually, the solubility product must be much less since all the linear plots for the dependence of the maximum stripping current on the I^- concentration arise from the coordinate origin. In those cases in which the sensitivity of the determination is limited by the solubility of the compound formed, this phenomenon is not observed /92/.

Elementary iodine may easily be determined after prior reduction by hydrazine in the cold. The determination is not disturbed by elements which do not precipitate with rhodamine S. Bismuth, iron, zinc, nickel, indium, lead, copper, and mercury may be present in a large excess. The determination of iodide ions is disturbed by antimony (III) which is concentrated on the electrode simultaneously with the iodine. The effect of antimony may be eliminated by reducing it to the elementary state or binding it in an electrochemically inactive complex. Thus, in the presence of 10% potassium citrate, a hundredfold excess of antimony (III) does not interfere with the determination of iodide ions (Sb^{3+} and I^- concentrations are $6 \cdot 10^{-5}$ and $6 \cdot 10^{-7}$ g-ion/liter). The maximum current for the electrochemical dissolution of a deposit containing iodine is somewhat lower in this case than in the absence of citrate ions. When a calibration graph or known addition is used, this behavior has no significance.

Techniques have been developed for the determination of iodine in halide salts and in sea water employing graphite (type I) and saturated calomel electrodes.

The determination is based on the reaction

$$2I^- + Cl^- + R^+ \rightleftharpoons R[I_2Cl] \downarrow + 2e$$

Determination of iodine in halide salts /106/. First, a 2 g salt sample is dissolved in 30 ml 0.1 M sulfuric acid, 10 mg hydrazine chloride is added, the solution is stirred, 0.5 ml 0.2% rhodamine S is introduced, and the solution volume is brought up to 40 ml with sulfuric acid. The solution is transferred to the electrolytic cell. The electrolysis of the stirred solution is carried out for 10 min at $+0.8$ V and iodine is concentrated on the electrode as a $R[I_2Cl]$ precipitate. The stirring is stopped and the cathodic polarization curve is recorded at from 0.8 to 0 V. The maximum cathodic current is measured. The iodine concentration is found from a known addition. A standard solution is prepared by dilution of potassium iodide recrystallized from alcohol.

In the determination of $2 \cdot 10^{-6}\%$ iodine, the coefficient of variation is 15%.

This method may be used to analyze other salts and complicated mixtures. In these cases, an adequate amount of chloride ions should be specially introduced into the solution.

Determination of iodine in sea water /106/. First, 10 mg hydrazine chloride is introduced into 30 ml sea water, the solution is mixed, 0.5 ml of 0.2% rhodamine S is added and the solution brought up to 40 ml with 0.2 N hydrochloric acid. The solution is transferred to the electrolytic cell and the analysis is continued as described in the previous technique.

The minimum determinable concentration is $6 \cdot 10^{-9}$ g-ion/liter.

In the determination of $1 \cdot 10^{-7}\%$ iodine, the coefficient of variation is 20%.

This method may be used to determine iodide ions in the presence of elementary iodine by adding hydrazine to eliminate iodine reduction.

Cobalt

Concentrating cobalt as compounds with nitrosonaphthols

Cobalt forms insoluble compounds with nitrosonaphthols. The solubility of the cobalt (III) compounds is somewhat lower than that of cobalt (II) compounds which allows concentrating cobalt (II) in solution in the form of an insoluble cobalt (III) compound. Since nitrosonaphthols are selective and the electrode reaction $Co^{2+} \rightleftarrows Co^{3+} + e$ occurs in a convenient potential range, the method is quite specific /104/.

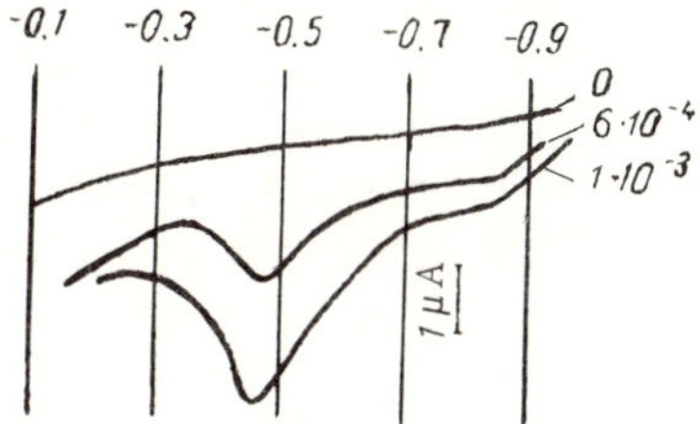

FIGURE 39. Polarization curves for the oxidation of cobalt (II) on the graphite electrode in solutions 0.4 M in NH₄OH and 0.05 M in NH₄Cl at different Co^{2+} concentrations (numbers on the curves: g-ion/liter).

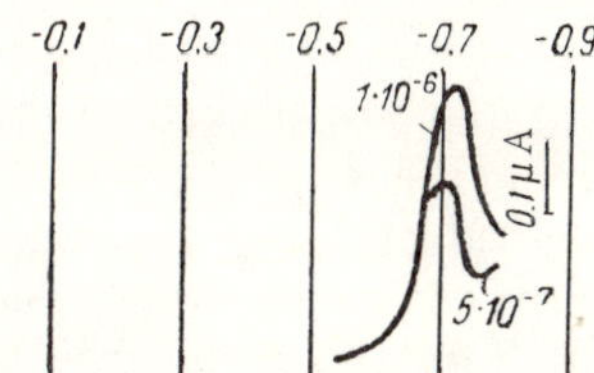

FIGURE 40. Polarization curves for the reduction ($v = 0.011$ V/sec) of the cobalt (III) compound with 2-nitroso-1-naphthol (RH) deposited at $\varphi_{el} = -0.5$ V and $\tau_1 = 1$ min from solutions 0.4 M in NH₄OH, 0.05 M in NH₄Cl, and $1.5 \cdot 10^{-4}$ M in RH with different Co^{2+} concentrations (numbers on the curves: g-ion/liter).

Figure 39 shows the anodic polarization curves for the electrochemical oxidation of Co^{2+} on the graphite electrode in a supporting electrolyte solution 0.4 M in NH₄OH and 0.05 M in NH₄Cl. An increase in the cobalt ion concentration is accompanied by a rise in the anodic current. The cathodic polarogram

recorded after anodic polarization of the electrode in such a
solution does not have maxima characteristic of the reduction
of compounds localized on the electrode surface.   Maxima
appear only after the introduction of 2-nitroso-1-naphthol
(Figure 40) or 1-nitroso-2-naphthol.*

Evidently cobalt is concentrated on the electrode as a result
of the following reactions:

$$Co^{2+} \longrightarrow Co^{3+} + e$$
$$Co^{3+} + 3RH \longrightarrow CoR_3 \downarrow + 3H^+$$

where RH is 1-nitroso-2-naphthol or 2-nitroso-1-naphthol.

The electrode reaction probably involves complex cobalt
ions or ions containing cobalt (II) with nitrosonaphthol.

An increase in the reagent concentration is accompanied by
a rise in the maximum cathodic current for the reduction of the
compound which forms in the concentrating step.   However,
introduction into the solution of more than $1.2 \cdot 10^{-3}\%$ nitrosonaphthol is unwise since this leads to distortion of the curve
and measuring the current becomes difficult.   In the determi-
nation of Co^{2+} in concentrations of $1 \cdot 10^{-7}$ to $1 \cdot 10^{-8}$ g-ion/liter,
the optimal reagent concentration should be $1 \cdot 10^{-5}—4 \cdot 10^{-5}\%$
(about $2 \cdot 10^{-6}$ M).

An electrolysis potential at from -0.5 to 0 V does not affect
the maximum reduction current of the compound which forms on
the electrode.   This potential range coincides with region of the
diffusion branch of the oxidation polarization curve for cobalt
ions (Figure 39).

The reduction current of the reagent observed at from -0.6
to -0.7 V does not interfere with the recording of the polarization
curve for the reduction of cobalt, which has a maximum at about
-0.8 V in an alkaline medium.   In a neutral medium, with an
0.5 M supporting solution of ammonium chloride, the oxidation-
reduction of cobalt and the reduction of the reagent occur at
more positive potentials.   The maximum cathodic current of
2-nitroso-1-naphthol is observed at from 0.1—0 V, and the
maximum current for the electrochemical reduction of cobalt
at about -0.1 V.   A short polarization of the electrode at a
potential preceding the dissolution of the cobalt-nitrosonaphthol

* To avoid the chemical oxidation of Co^{2+} ions, the cobalt salt or the reagent is introduced
into the solution after the removal of oxygen. Such solutions containing $10^{-8}—10^{-7}$ g-ion/
/liter Co^{2+} and $2 \cdot 10^{-6}$ M reagent are quite stable over long periods of time.

compound (-0.5 V in an alkaline medium and 0 V in a neutral medium) results in the vanishing of the reduction current of the reagent before the beginning of the reduction of cobalt. The polarogram for the reduction of the cobalt compound recorded after such treatment is well defined.

Currents for the electrochemical reduction of compounds of cobalt (III) with 2-nitroso-1-naphthol at the same Co^{2+} ion concentration in the solution are approximately twice higher than the maximum currents for the electrochemical reduction of 1-nitroso-2-naphthol compounds of cobalt. At low reagent concentrations, this phenomenon is not observed.

When the electrochemical deposition lasts 5 min, a directly proportional relation is observed between the maximum reduction current and the cobalt ion concentration in the $0-8 \cdot 10^{-7}$ g-ion/liter range. At higher Co^{2+} ion concentrations, a saturation region appears on the $i_{max} = f(c)$ curve apparently due to the formation of insoluble compounds of Co^{2+} with the reagents in the solution bulk. The dependence of the maximum current for the electrochemical reduction of the compound formed on the electrode on the concentrating step time is directly proportional at Co^{2+} ion concentrations of about 10^{-7} g-ion/liter and has a complicated nonlinear character at concentrations of about $1 \cdot 10^{-6}$ g-ion/liter. This phenomenon is probably due to the fact that the expanding film of the compound passivates the electrode. The complicated nature of the dependence of the maximum current on the electrolysis time does not prevent us from using this technique for concentrating and determining microgram amounts of cobalt, since we can easily select an electrolysis time for the required ion concentration range for which the current is directly proportional on the concentration. The method may be used for the determination of $1 \cdot 10^{-8} - 1 \cdot 10^{-6}$ g-ion/liter Co^{2+}.

The determination of cobalt is not disturbed by elements which do not react with nitrosonaphthols and are electrochemically inactive at from -0.5 to -0.8 V relative to the saturated calomel electrode. Other elements may be present in fairly large amounts. Copper, iron, and zinc do not interfere with the determination of cobalt in thousandfold excesses, nor does nickel in concentrations up to $1.2 \cdot 10^{-4}$ g-ion/liter. When the nickel concentration is higher (up to 0.01 g-ion/liter), the reagent concentration should be increased. Tin (IV), titanium, and chromium (III) may be present in the solution in a tenfold excess

with respect to the cobalt, and silver and chromium (VI)
in a hundredfold excess.

Techniques have been developed for the determination of
cobalt in neodymium molybdate, yttrium molybdate, erbium
tungstate /104/, calcium chloride, citric acid, tartaric acid, cadmium nitrate, and nickel carbonate /130/ employing graphite
(type I) and saturated calomel electrodes. The determination of
cobalt is based on the reaction

$$Co^{2+} + 3RH^* \rightleftarrows CoR_3 \downarrow\ + 3H^+ + e$$

*Determination of cobalt in neodymium molybdate, yttrium
molybdate, and erbium molybdate* /104/. The difficulties in the
polarographic analysis of the molybdates and tungstates of neodymium, yttrium, and erbium are due to the electrochemical
activity of molybdate and tungstate ions in acid solution and the
hydrolysis of the rare earth elements in alkaline medium. In
order to eliminate these undesirable phenomena, the samples
are dissolved in hydrochloric acid (1:1), then citric acid (to
complex the erbium, neodymium, and yttrium present) and sodium
hydroxide are added. During the neutralization, an acid salt
precipitate forms which dissolves when the pH is increased. The
introduction of citrate ions is accompanied by a shift in the oxidation potential of Co^2 in the positive direction and a decrease
in the maximum cathodic current. In this respect, the conditions
for the analysis of erbium, yttrium, and neodymium molybdate
are somewhat different from the optimum conditions cited above
for ammonia-ammonium solutions.

First, an 0.35 g sample is dissolved in 2 ml hydrochloric acid
(1:1). The solution is evaporated somewhat, 5 ml 10% citric acid
is added, and a solution of sodium hydroxide is added to bring the
pH to 8. Then the volume of the mixture is brought up to 20 ml
with a solution 0.05 M in NH_4Cl and 0.4 M in NH_4OH. The solution is transferred to the electrolytic cell, nitrogen is passed for
10 min, and then 0.2 ml of a $2 \cdot 10^{-4}$ M alkaline solution of 2-nitro-
so-1-naphthol are added. The electrolysis of the stirred solution
is carried out for 5 min at 0 V and cobalt is concentrated on the
electrode as a nitrosonaphthol compound. The stirrer is stopped
and the cathodic polarization curve for the electrochemical

---

* RH is 2-nitroso-1-naphthol.

reduction of the cobalt-nitrosonaphthol compound is recorded from -0.5 to -0.8 V. The maximum cathodic current is measured and the Co^{2+} ion concentration is found from a calibration graph or from a known addition.

The sensitivity of the determination of cobalt in erbium, yttrium, and neodymium molybdate is $3 \cdot 10^{-5}\%$ for a 0.35 g sample. The coefficient of variation is 15%.

*Determination of cobalt in erbium tungstate, yttrium tungstate, and neodymium tungstate.** First, an 0.5 g sample is dissolved in 1 ml (in the case of neodymium, 2 ml) orthophosphoric acid with heating, and the solution is diluted to 20 ml with distilled water. Then, 2 ml of this solution are drawn off with a pipette, 1 ml (0.5 ml, in the case of erbium) 50% citric acid is added, then concentrated sodium hydroxide is added until the precipitate formed in the neutralization dissolves, and the volume of the mixture is brought up to 20 ml with a solution 0.4 M in NH_4OH and 0.05 M in NH_4Cl. The solution is transferred to the electrolytic cell, dissolved oxygen is removed by bubbling inert gas, and then 0.2 ml of a solution 1 N in NaOH and $2 \cdot 10^{-4}$ M in 2-nitroso-1-naphthol are introduced. The electrolysis of the stirred solution is carried out for 10 min (5 min, in the case of erbium) at a potential of 0 V and cobalt is concentrated on the electrode as a nitrosonaphthol compound. The stirring is stopped and the cathodic polarization curve for the reduction of the cobalt (III)--nitrosonaphthol compound is recorded at from -0.5 to -1.0 V. The value of the maximum cathodic current is measured. The cobalt concentration is found from a calibration graph or from a known addition.

In the determination of $5 \cdot 10^{-4}\%$ cobalt, the coefficient of variation does not exceed 10%.

*Determination of cobalt in calcium chloride.*** First, a 2 g calcium chloride sample is dissolved in 20 ml 0.1 M ammonium chloride, and the pH is brought up to 8.5 with ammonium hydroxide. The solution is transferred to the electrolytic cell, oxygen is removed by bubbling inert gas, and 0.2 ml of a solution $2 \cdot 10^{-4}$ M in 2-nitroso-1-naphthol and 0.1 N in NaOH are added. The electrolysis of the stirred solution is carried out for 15 min at 0 V and cobalt is concentrated on the electrode as nitrosonaphthol compound. Then the stirring is stopped and the cathodic polarization curve is recorded from -0.4 to -1.0 V. The

* The technique was proposed by T.A.Krapivkina and Kh.Z.Brainina.
** The technique was proposed by Kh.Z.Brainina and T.A.Krapivkina.

maximum cathodic current is measured and the cobalt concentration is found from a calibration graph or from a known addition.   Figures 41 and 42 show the polarization curves for the electrochemical reduction of cobalt (III) concentrated from a calcium chloride solution and the calibration graph.

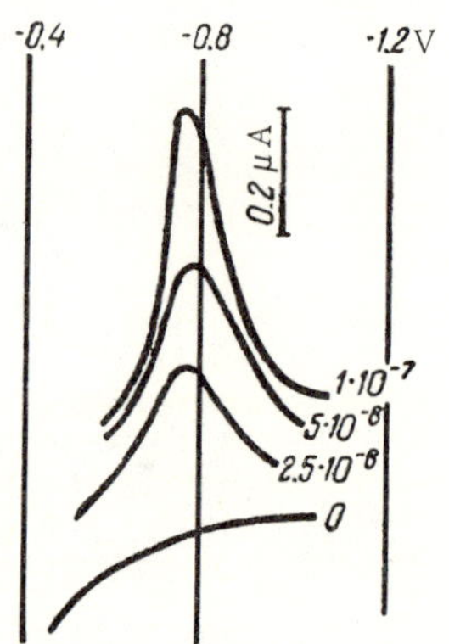

FIGURE 41. Polarization curves for the reduction of a cobalt (III) compound deposited at $\varphi_{el} = 0$ and $\tau_1 = 15$ min from solutions 1 M in $CaCl_2$, 0.1 M in NH_4Cl, and $2 \cdot 10^{-6}$ M in 2-nitroso-1-naphthol (pH = 8.5) with different Co (II) concentrations (numbers on the curve: g-ion/liter).

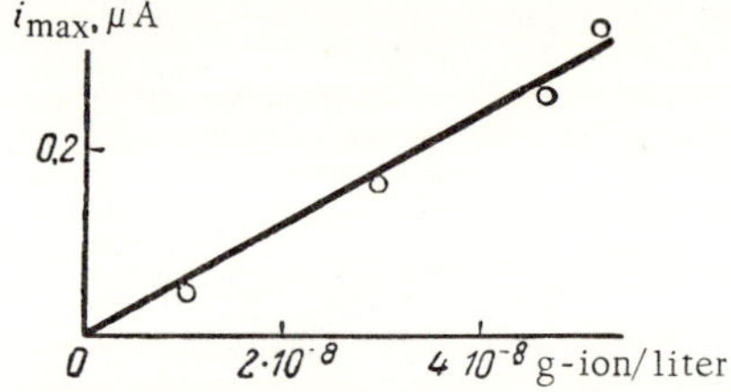

FIGURE 42. Dependence of the maximum reduction current for the cobalt (III)-2-nitroso-1-naphthol compound on the Co (II) concentration (for conditions see Figure 41).

In the determination of $1 \cdot 10^{-6}\%$ cobalt, the coefficient of variation is 10%.

*Determination of cobalt in citric and tartaric acids.** First, a 1 g acid sample is dissolved in 5 ml water, a sodium hydroxide solution is added until the pH is between 8.5 and 9.0 in the analysis of tartaric acid and between 10 and 11 in the analysis of

---

* The technique was proposed by Kh.Z.Brainina and T.A.Krapivkina.

citric acid, and then a solution 0.1 M in NH_4OH and 0.05 M in NH_4Cl is added to bring the total volume to 25 ml.   The solution is transferred to the electrolytic cell, oxygen is removed by bubbling inert gas, and 0.25 ml $2 \cdot 10^{-4}$ M 2-nitroso-1-naphthol is introduced.   The electrolysis of the stirred solution is carried out for 10 min at 0 V and cobalt is concentrated on the electrode as a coordination compound with 2-nitroso-1-naphthol.   The stirring is stopped and the cathodic polarization curve recorded at from -0.5 to -1.0 V.   The maximum cathodic current is measured.   The cobalt concentration is found from a known addition.

In the determination of $1.5 \cdot 10^{-5}\%$ cobalt, the coefficient of variation does not exceed 10%.

*Determination of cobalt in cadmium nitrate.** First, a 1 g cadmium nitrate sample is dissolved in 25 ml of a solution 1 M in NH_4OH and 0.1 M in NH_4Cl from which oxygen had previously been removed by bubbling inert gas, and 0.25 ml of a solution $2 \cdot 10^{-4}$ M in 2-nitroso-1-naphthol and 0.1 N in sodium hydroxide is introduced.   The electrolysis of the stirred solution is carried out for 5 min at -0.5 V.   Cobalt is concentrated on the electrode as coordination compound of Co (III) with 2-nitroso-1-naphthol. The stirring is stopped and the cathodic polarization curve is recorded at from -0.5 to -0.85 V.   The maximum cathodic current is measured, and the cobalt concentration is found from a known addition.

In the determination of $7 \cdot 10^{-6}\%$ cobalt, the coefficient of variation is 10%.

Determination of cobalt in basic nickel carbonate /130/. First, an 0.5 g nickel carbonate solution is dissolved in 2 ml hydrochloric acid, the solution is diluted to 20 ml with water and transferred to a separating funnel.   Then, 5 ml 5 M ammonium thiocyanate and 0.2 ml diantipyrylmethane (DM) in 0.5 M hydrochloric acid are added.   The cobalt is extracted with 2 ml chloroform, 0.3 ml portions of ammonium thiocyanate and DM are added, and the extraction is repeated.   The organic fractions are collected in a porcelain crucible, evaporated to dryness over a water bath, and calcined in a muffle furnace for 1 hr at 600°C.   After the crucible cools, 0.5 ml concentrated hydrochloric acid is added to the dry residue, the walls of the crucible are carefully washed with slight heating on an electric hot plate

---

* The technique was proposed by T.A.Krapivkina and Kh.Z.Brainina.

until the residue dissolves completely, and a deaerated solution 1.5 M in NH₄OH and 0.1 M in NH₄Cl is added up to 25 ml. The solution is transferred to the electrolytic cell. Oxygen is removed by bubbling inert gas, and 0.5 ml of a solution $1 \cdot 10^{-4}$ M in 2-nitroso-1-naphthol and 0.1 N in sodium hydroxide is introduced. The electrolysis of the stirred solution is carried out for 5 min at -0.45 V and cobalt is concentrated on the electrode as a nitrosonaphthol compound. The stirring is stopped. After the solution settles, the electrode potential is shifted to -0.55 V, and the cathodic polarization curve is recorded at from -0.55 to -0.90 V. The maximum cathodic current is recorded ($\varphi_{max} = -0.75$ V). The cobalt concentration is found from a known addition.

The sensitivity of the determination is $1 \cdot 10^{-5}\%$.

The extraction may be omitted if a sensitivity of $2 \cdot 10^{-4}\%$ is adequate. The determination of $5 \cdot 10^{-8}$ g-ion/liter Co^{2+} is not disturbed by 0.4 g-ion/liter Ni^{2+}. In this case, the 2-nitroso-1-naphthol concentration in the sample solution should be increased to $2 \cdot 10^{-5}$ M.

Concentrating cobalt as cobalt diethyl-dithiocarbamate /131/

Cobalt forms an insoluble compound with diethyldithiocarbamate ions. The solubility of the cobalt (II) compound is higher than the solubility of the cobalt (III) compound, thereby facilitating the concentrating of the oxidized form of the element on the electrode. Sodium diethyldithiocarbamate (DDTC) has been studied as a reagent for extractions, precipitations /132, 133/, and amperometric titrations /134—137/. DDTC forms insoluble compounds with the ions of many metals and is not selective or sensitive. These drawbacks are eliminated to some extent, if DDTC is used as a precipitating agent in electrochemical reactions. The selectivity of the determination in this case is achieved due to the individual electrochemical properties of the element being determined and high sensitivity is obtained by concentrating the reaction product on the electrode surface.

Figure 43 presents the polarization curves for the oxidation of DDTC and further electrochemical conversions of the electrode reaction products.* The electrochemical oxidation of

---

* Solutions of DDTC (1 g in 100 ml water) should be freshly prepared and introduced after removal of oxygen.

DDTC begins at -0.2 V. The diffusion current is observed in
the 0—0.4 V range. Usatenko and Tulyupa /135/ detected an
anodic wave for the reagent at a more positive potential. This
discrepancy is apparently due to different acidities of the solutions studied. The same authors found that the oxidation occurs
in two one-electron steps, in the first of which an insoluble disulfide forms. Figure 43 shows that the insoluble product of the
anodic reaction (a) is concentrated on the electrode and undergoes
further cathodic (b) and anodic (c) electrochemical conversions.
The deposit forms in the electrolysis at potentials between 0 and
0.2 V, which correspond to the limiting diffusion current of DDTC.
The maximum cathodic current (and the amount of deposit on the
electrode) increase with increasing deposition time. The polarization curves for the oxidation of the disulfide are similar. At
negative potentials, the electrode deposit virtually does not form.
The optimal range for the concentrating step is 0—0.2 V; the
following electrode reaction, which is accompanied by dissolution
of the deposit, becomes appreciable at more positive potentials.

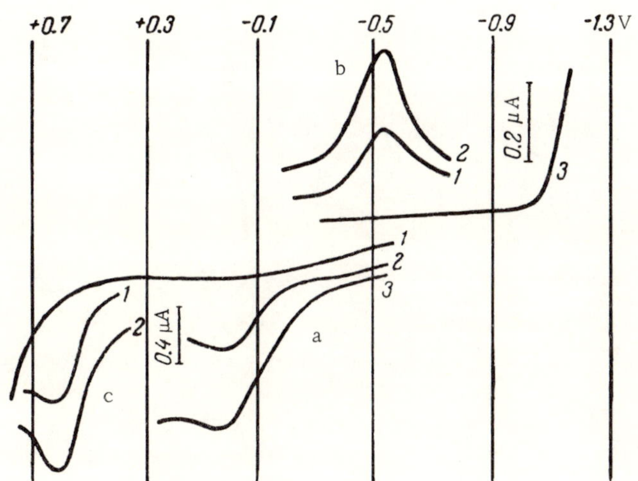

FIGURE 43. Polarization curves for the oxidation of sodium diethyldithiocarbamate (a),
and the reduction (b) and oxidation (c) of the first electrode reaction products deposited
from solutions 0.1 M in NH_4OH, 0.1 M in NH_4Cl, and containing different amounts of
sodium diethyldithiocarbamate (n):

a: 1) $n = 0$ M, 2) $n = 1.7 \cdot 10^{-4}$ M, 3) $n = 3.4 \cdot 10^{-4}$ M;

b: $\varphi_{el} = 0$ V, 1) $n = 1.2 \cdot 10^{-5}$ M, $\tau_1 = 2$ min, 2) $n = 2.4 \cdot 10^{-5}$ M, $\tau_1 = 2$ min,
3) $n = 0$, $\tau_1 = 0$ min;

c: $\varphi_{el} = +0.2$ V, $\tau_1 = 5$ min, 1) $n = 8 \cdot 10^{-7}$ M, 2) $n = 1.5 \cdot 10^{-6}$ M.

The limiting diffusion current for the electrochemical oxidation of DDTC is directly proportional to its concentration. A similar relationship exists between the maximum current for the oxidation or reduction of the disulfide formed on the electrode and the DDTC concentration. Thus, concentrating the reagent on the electrode as an insoluble oxidation product is feasible.

Information concerning the reagent concentration or the amount of electrode deposit may be obtained by recording the anodic or cathodic polarization curves for the electrochemical dissolution of the compound. Very small amounts of the reagent in the solution may be determined. Indirect determination of those elements which form insoluble compounds with diethyldithiocarbamate is feasible.

The electrochemical stability of DDTC at potentials more negative than -0.2 V allows use of the reagent to concentrate the metal ions which are electroactive at these potentials as insoluble diethyldithiocarbamates. For example, cobalt ions may be concentrated and determined by running the electrochemical reaction in an ammonia-ammonium medium.

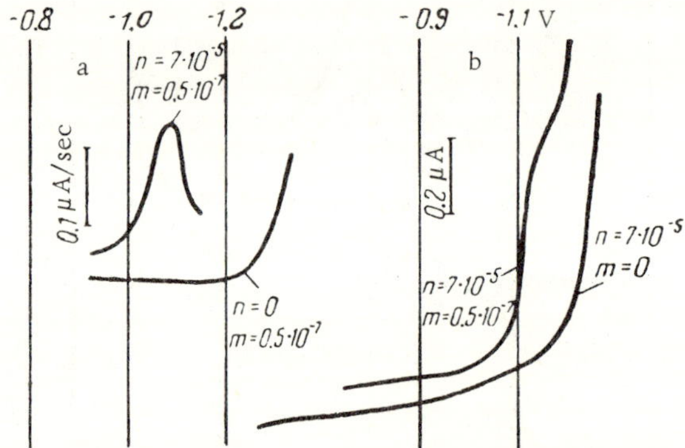

FIGURE 44. Differential (a) and integral (b) polarization curves for the reduction ($v = 0.011$ V/sec) of cobalt (III) diethyldithiocarbamate deposited for $\varphi_{el} = -0.5$ V and $\tau_1 = 5$ min from solutions 0.1 M in NH$_4$OH, 0.1 M in NH$_4$Cl, and having different values of n (sodium diethyldithiocarbamate content, M) and m (Co^{2+} content, g-ion/liter).

Polarization curves recorded after the electrolysis of a number of solutions are presented in Figure 44. A cathodic

current is seen to arise only when the solution contains both
the reagent and Co^{2+}. Evidently, Co^{2+} is concentrated in the
anodic polarization of the electrode in an ammonia-ammonium
buffer solution containing sodium diethyldithiocarbamate as
cobalt (III) diethyldithiocarbamate. The optimal potential for
the concentrating step is -0.5 V. The electrochemical reduction
of the compound occurs at from -1.0 to -1.2 V but is masked on
the integral curve by the reduction current of the background.
The differential curves are more pronounced and more suitable
for recording. The reagent concentration influences the shape
of the curves. The concentration should be $5 \cdot 10^{-5}$—10^{-6} M.

The maximum current for the electrochemical reduction or
its derivative with respect to time) are directly proportional on
the Co^{2+} ion concentration and the deposition time. The maximum current increases as the potential scanning rate increases.

Well-defined differential polarograms are recorded after
concentrating cobalt for 5 min from a solution 0.1 M in NH_4Cl,
0.1 M in NH_4OH, $7 \cdot 10^{-5}$ M in DDTC, and $1 \cdot 10^{-7}$ M in Co (II).

The relative mean square error of the determinations is 10%.

Accordingly, sodium diethyldithiocarbamate may be used as
a precipitating agent in the cathodic stripping voltammetry of
cobalt.

The determination of cobalt is not disturbed by chromium (VI)
or zinc (II) in thousandfold excesses, or nickel (II) or cadmium
(II) in hundredfold excesses. Elements which do not interact
with DDTC and are not electrochemically active at potentials
ranging from -0.2 to -1.2 V do not interfere in any amounts.

Nickel

Complexes of nickel (II) and (IV) with dimethylglyoxime are
known. The ability of nickel (II) ions to undergo oxidation and
the stability of its complexes is largely dependent on the alkalinity of the medium /138, 139/. Nickel is presumably capable
of possessing a valence of three. In alkaline solutions containing nickel (II) and dimethylglyoxime electrochemical oxidation of complex metal-dimethylglyoxime ions and free dimethylglyoxime is observed. The anodic polarograms of nickel
and dimethylglyoxime are presented in Figure 45. In an alkaline

solution not containing dimethylglyoxime, the half-wave potential
of nickel is about $+0.45$ V (curve 1). When the reagent is introduced, the anodic wave is shifted toward more negative potentials
though its intensity remains virtually unchanged (curve 2).
Further addition of dimethylglyoxime results in the appearance
of a second anodic wave at more positive potentials. Evidently,
the wave at the more negative potentials is due to the oxidation
of the nickel-dimethylglyoxime complex and the second wave,
to the oxidation of free reagent (curves 5—7). Curves 6 and 7
have waves at the oxidation potentials of nickel which increase
with increasing nickel concentration in the solution. The oxidation wave of dimethylglyoxime decreases. The reagent bound
in the complex evidently does not undergo any electrochemical
conversion.

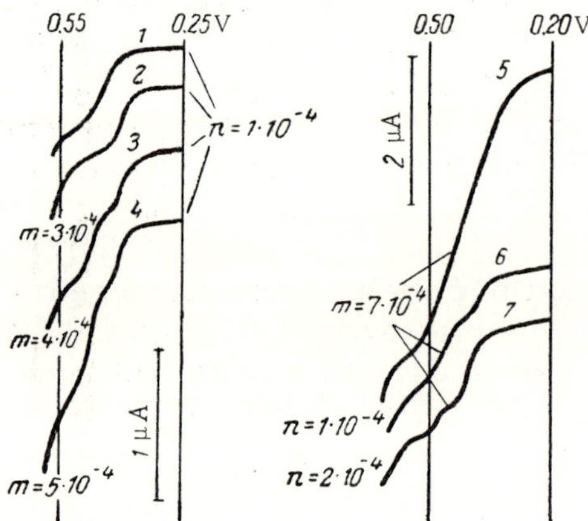

FIGURE 45. Polarization curves for the oxidation of the nickel-dimethylglyoxime complex and of dimethylglyoxime obtained from
the electrolysis of solutions 3 N in KOH with the dimethylglyoxime
concentrations m, M and Ni^{2+} concentrations n, g-ion/liter.

The destruction of the nickel (II) hydroxyl complex and the
formation of the nickel (II) complex with dimethylglyoxime after
the addition of dimethylglyoxime to the solution occur slowly.
At first, the total oxidation wave of both components is observed
and then gradually decreases. At more negative potentials
a new wave appears corresponding to the oxidation of nickel
dimethylglyoximate. After 20—30 min, the curve takes the form
of curve 2. The oxidation current of dimethylglyoxime (H_2D)
arises when the ratio $H_2D:Ni$ is greater than 3 and the oxidation

current of the nickel complex with dimethylglyoxime appears
when the ratio is 1:1. Thus, the maximum number of ligands
in the nickel (II) complex with dimethylglyoxime is apparently
three, and the minimum, one. Okâč and Širnek /143/ arrived
at a similar conclusion on the basis of photometric titration
data, and Devis and Boudreaux /138/ supported this finding in
a polarographic and magnetic study.

In the electrolysis of alkaline solutions containing Ni^{2+} ions
($1 \cdot 10^{-5}$—$3 \cdot 10^{-5}$ g-ion/liter) and dimethylglyoxime ($5 \cdot 10^{-5}$—
—$1 \cdot 10^{-4}$ M) the electrode becomes coated with a red deposit at
0.8 V. Coulometric and chemical analyses of the deposit show
that it consists of nickel (III) and dimethylglyoxime in a 1:1 ratio.
The electrochemical formation of the deposit on the electrode
may be tentatively represented by the equation

$$Ni(HD)_3^- + 2OH^- \longrightarrow Ni(OH)_2HD \downarrow + 2HD^- + e$$

The number of hydroxyl groups or oxygens in the structure of
the compound is unknown. The compound also forms on the
electrode as a result of the reduction of the nickel (IV) complex
described by Devis and Boudreaux /138/. Both reactions may be
used to concentrate and determine microgram amounts of nickel.
Only the first methods involving oxidation and reduction of the
compound formed in a cycle has as yet been studied extensively.

The cathodic polarization curves obtained after the electrolysis of 0.005 M hydroxide potassium containing varying amounts
of dimethylglyoxime and nickel ions at different electrode potentials are presented in Figure 46. The anodic background
oxidation current, which has no significant point of inflection,
becomes the ascending branch of the cathodic curve, which has a
well-defined maximum. This maximum may easily be measured
using the right-hand portion of the curve. The maximum current for the electrochemical reduction of the deposit is a function
of the electrode potential in the concentrating step. In the
electrolysis of dilute solutions of nickel sulfate ($2 \cdot 10^{-7}$M), the
maximum cathodic current increases when the electrode potential in the electrolysis is shifted in the anodic direction to
0.8 V and then remains constant. In the electrolysis of solutions with Ni^{2+} concentrations of $3 \cdot 10^{-5}$ g-ion/liter, the maximum current is not dependent on the electrode potential in the
anodic cycle starting from 0.65 V.

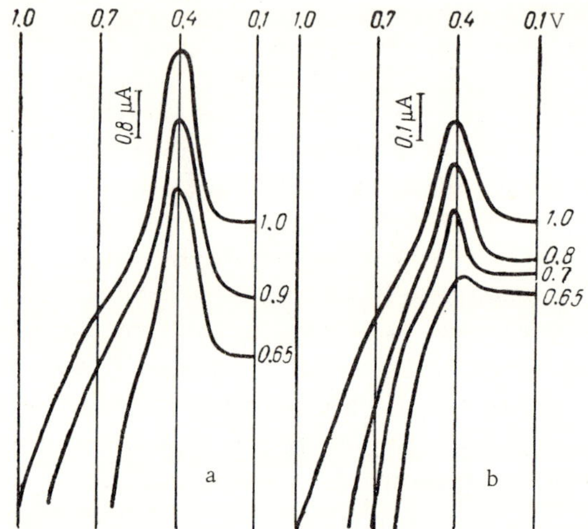

FIGURE 46. Polarization curves for the reduction of a precipitate deposited at different φ_{el} (numbers on the curves) from solutions 0.05 M in KOH with various Ni^{2+} and dimethylglyoxime concentrations:

a) $[Ni^{2+}] = 3 \cdot 10^{-5}$ g-ion/liter, $[DH_2] = 1 \cdot 10^{-4}$ M, $\tau_1 = 3$ min; b) $[Ni^{2+}] = 2 \cdot 10^{-7}$ g-ion/liter, $[DH_2] = 5 \cdot 10^{-6}$ M, $\tau_1 = 3$ min.

The nature of the polarization curves depends on the hydroxyl ion concentration in the solution. In slightly alkaline solutions, the electrochemical reduction current is poorly defined, apparently due to the binding of nickel (II) ions by dimethylglyoxime in an insoluble compound in the solution bulk. When the alkalinity is increased to $5 \cdot 10^{-2}$ M, the shape of the curve improves and the maximum cathodic current increases. When the potassium hydroxide concentration is further increased, the maximum current falls and the area bounded by the polarization curve and the residual current line decreases, indicating a deterioration in the concentrating step conditions. This is probably caused by the formation of a soluble compound between the nickel ions resulting from the anodic process with dimethylglyoxime in slightly alkaline medium which is not concentrated on the electrode. Nevertheless, the nickel-dimethylglyoxime compound may be obtained on the electrode even from more alkaline solutions if the electrolysis potential is shifted to a more positive region along with the increase in alkalinity. Under these conditions, the discharge of hydroxyl ions resulting in acidification of the layer adjacent to the electrode occurs along with the oxidation of nickel (II) on the electrode. The intensity of this reaction, which we can judge from the electrolysis current, increases with increasing electrode potential, compensating for the unfavorable conditions in the medium. Small amounts of nickel are conveniently concentrated at 0.8 V from 0.02—0.06 M potassium hydroxide. More alkaline solutions should be used for relatively high nickel concentrations ($\geqslant 5 \cdot 10^{-6}$ g-ion/liter) and in the analysis of amphoteric metal compounds.

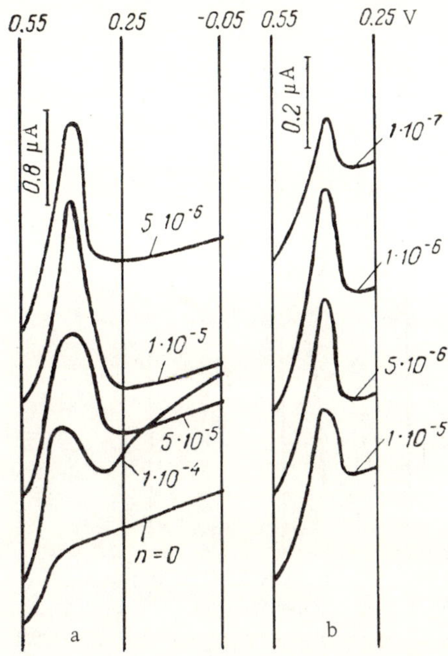

FIGURE 47. Polarization curves for the dissolution of precipitates deposited at $\varphi_{el} = 0.8$ V from solutions 0.04 M in KOH and n M in DH_2 (numbers on the curve):

a) $[Ni^{2+}] = 1 \cdot 10^{-6}$ g-ion/liter, $\tau_1 = 15$ min; b) $[Ni^{2+}] = 2 \cdot 10^{-7}$ g-ion/liter, $\tau_1 = 10$ min.

The effect of the precipitating agent concentration on the nature of the polarization curves for the electrochemical reduction of a precipitate is illustrated in Figure 47. In the absence of dimethylglyoxime, the concentrating of nickel is not observed.

When the reagent concentration is high, an additional process due to the oxidation of dimethylglyoxime is recorded after the current for the electrochemical reduction of the deposit. A curve form convenient for measuring is obtained when a solution containing a ten- to hundredfold excess of the reagent with respect to the nickel ions is used. Reasonable relations between the nickel (II) ion and dimethylglyoxime concentrations in solution are:

$c_{Ni^{2+}}$, g-ion/liter ...	$2 \cdot 10^{-9}$	$5 \cdot 10^{-9}$	$1 \cdot 10^{-8}$	$5 \cdot 10^{-8}$	$1 \cdot 10^{-7}$	$5 \cdot 10^{-7}$	$1 \cdot 10^{-6}$
c_{DH_2}	$1 \cdot 10^{-6}$	$1 \cdot 10^{-6}$	$1 \cdot 10^{-6}$	$2 \cdot 10^{-6}$	$5 \cdot 10^{-6}$	$5 \cdot 10^{-6}$	$1 \cdot 10^{-5}$

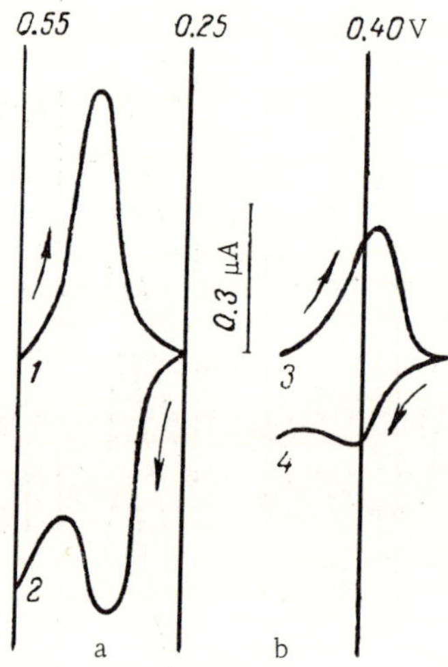

FIGURE 48. Cyclic polarization curves for the reduction (1, 3) and oxidation (2, 4) of a nickel-dimethylglyoxime compound obtained from the electrolysis (φ_{el} = 1.0 V, τ_1 = 10 min) of a solution 0.1 N in KOH and $1 \cdot 10^{-5}$ M in DH_2:

a) $[Ni^{2+}]$ = $1 \cdot 10^{-6}$ g-ion/liter; b) $[Ni^{2+}]$ = $5 \cdot 10^{-7}$ g-ion/liter.

An interesting feature of this system is the possibility of reproducing the cathodic polarization curve many times after a short polarization of the electrode at 0.8 V (without lengthy electrolysis after recording the cathodic polarogram). The cyclic polarograms (Figure 48) show oxidation-reduction currents for the compound formed a single time on the electrode. Evidently, the deposit accumulated on the electrode in the concentrating step is reduced and converted to an insoluble nickel (II) compound which remains on the electrode rather than being dissolved when the electrode potential is shifted in the negative direction. Apparently, the electrochemical reaction occurs in the solid phase. Moreover, the nickel (II)-dimethylglyoxime compound in the alkaline medium does not spontaneously precipitate but when the precipitate is formed under other conditions, it does not dissolve in the alkaline medium. The oxidation-reduction effect without stripping of the compounds formed is evidently due to the properties of the nickel (II)-dimethylglyoxime compound. The deposit may be removed from the electrode surface mechanically or chemically (e. g., by dissolution in 0.1 N sulfuric acid).

Another feature of the nickel-dimethylglyoxime system is the dependence of the maximum current for the electrochemical reduction of the compound on the formation time from freshly prepared solutions. The maximum cathodic current increases linearly with increasing deposition time. The electrode is not passivated over a wide concentration and time range. The linear plots describing the dependence of the maximum reduction current on the prior electrolysis time cut off increasing segments on the abscissa with slower precipitating-agent concentration. At a fairly high dimethylglyoxime concentration, the linear plot intersects the coordinate origin. This phenomenon is explained by the slow formation of electroactive nickel-dimethylglyoxime complexes (35—45 min). If the compound is concentrated from a solution which has stood for some time, the dependence of the maximum current for the electrochemical reduction of the compound on its formation time has the usual form. This phenomenon should be taken into account by using a known addition. The standard solutions introduced into the sample solution should contain complexes of nickel (II) and dimethylglyoxime. Usually 3 N potassium hydroxide solutions containing nickel salts and dimethylglyoxime are used. The ratio of nickel and dimethylglyoxime is chosen according to the data given above.

Polarization curves obtained after the electrolysis of solutions with an Ni^{2+} concentration of $5 \cdot 10^{-9}$ g-ion/liter employing different graphite electrodes are given in Figure 49. The residual current in the potential range used is significantly higher with an electrode impregnated with an epoxy resin (type I) than with an electrode impregnated with a mixture of paraffin and polyethylene (type III).   The first type of electrodes should therefore not be used to determine ultrasmall amounts of nickel, while characteristic current maxima for the electrochemical reduction of nickel concentrated from solutions containing $2 \cdot 10^{-9}$ to $5 \cdot 10^{-9}$ g-ion/liter Ni^{2+} are observed when type III electrodes are used.

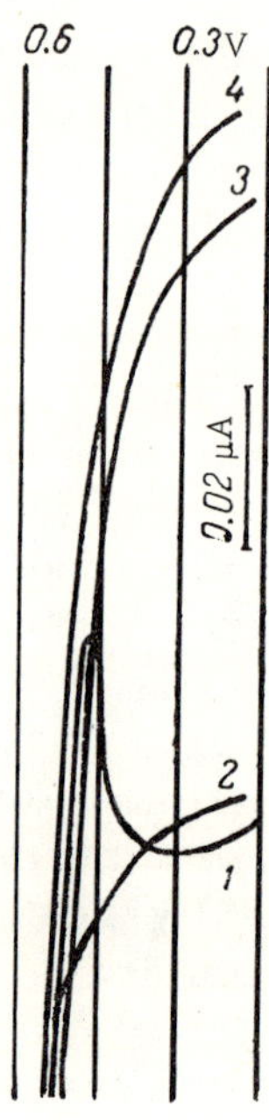

FIGURE 49. Polarization curves (1, 4) and residual current curves (2, 3) characterizing the reduction of precipitates deposited when $\tau_1 = 90$ min from a solution 0.05 M in KOH, $5 \cdot 10^{-6}$ M in DH_2, and $5 \cdot 10^{-9}$ g-ion/liter in Ni^{2+}:

1, 2) with type III graphite electrodes; 3, 4) type I.

The clearest curve is recorded when the voltage scan in the 0.8—0.2 V range is repeated.   This method may be used since the deposit of the compound once formed on the electrode does not dissolve in the cathodic cycle (Figure 48).

The following are the optimum conditions for concentrating and determining nickel: an 0.02—0.1 M KOH solution containing a ten- to hundredfold excess of dimethylglyoxime with respect to the nickel ions, electrolysis at an anode potential of 0.8—1.0 V relative to the saturated calomel electrode, washing the electrode in 0.1 N sulfuric acid after each measurement, using a type III electrode, and recording the curve twice.   Under these conditions, the cathodic current maximum is directly proportional to the Ni^{2+} concentration in the $2 \cdot 10^{-9}$ to $1 \cdot 10^{-6}$ g-ion/liter range.   The electrolysis time for solutions with Ni^{2+} concentrations of $2 \cdot 10^{-9}$ to 10^{-8} g-ion/liter is 1—1.5 hr.

The error in a single determination of 5 ng nickel in 20 ml solution ($\sim 5 \cdot 10^{-9}$ M) does not exceed 20% with a confidence probability of 0.95.

The determination of nickel is disturbed by elements which are electroactive in the working potential range such as manganese (II), chromium (III), and lead.   Concentrations of these metals should not exceed $1 \cdot 10^{-4}$ g-ion/liter.   The interfering effect of ions which hydrolyze at concentrations equal to or greater than $1 \cdot 10^{-4}$ g-ion/liter is eliminated here as in chemical methods by introducing nitrate ions to a concentration not exceeding 2%.   Tartrates are unsuitable because of their electrochemical oxidation (the anodic wave of tartrate ions is observed at $+ 0.35$ V).   Nickel may be determined in the presence of a thousandfold excess of cobalt, copper, or iron.

Techniques have been proposed for the determination of nickel in ammonium molybdate, sodium tungstate, aluminum nitrate /107/, citric acid, tartaric acid ($1 \cdot 10^{-6}$%), and margarin ($1 \cdot 10^{-5}$%) employing graphite (type III) and saturated calomel electrodes.

*Determination of nickel in ammonium molybdate and sodium tungstate.** Standard solutions of nickel with an Ni^{2+} concentration of 50 and 5 μg/liter are prepared by dilution of the nickel salt solution with 3 N potassium hydroxide solutions containing 0.5 mg/liter and 0.05 mg/ml dimethylglyoxime, respectively.   A solution may be used for two hours after preparation.

First, a 1 g calcined ammonium molybdate (or sodium tungstate) sample is weighed, 20 ml 5% (1% for sodium tungstate) base solution is added, followed by 0.2 ml solution 0.001% in dimethylglyoxime and 3 N in potassium hydroxide.   The resulting solution is allowed to stand for 1 hr.   The solution is then transferred to the electrolytic cell and the electrolysis of the stirred solution is carried out for 10 min at an indicator electrode poential of 0.8 V.   Nickel is concentrated on the electrode as a dimethylglyoxime compound.   The stirring is stopped.   The scanning of the voltage in the 0.8—0.2 V range is turned on, a potential of 0.8 V is established, and the cathodic polarization curve for the electrochemical reduction of the deposit is recorded at once. The maximum current observed on the polarogram in the

---

* The technique was proposed by E.Ya.Sapozhnikova.

0.50—0.35 V potential range is measured. The nickel concentration is found using a known addition. After each measurement the electrode surface is renewed mechanically or washed with 0.1 N sulfuric acid.

In the determination of $1 \cdot 10^{-5}\%$ nickel, the coefficient of variation does not exceed 15%. Increasing the deposition time to 45—60 min enables us to determine $2 \cdot 10^{-6}\%$ nickel with the same error.

*Determination of nickel in aluminum nitrate.** First, an 0.2 g salt sample is dissolved in 20 ml 0.5 N potassium hydroxide, 0.1 ml of a solution 0.001% in dimethylglyoxime and 3 N in potassium hydroxide is added, and the solution is allowed to stand for 1 hr. It is then transferred to the electrolytic cell and the electrolysis of the stirred solution is carried out for 30 min at 1 V. The analysis is continued as outlined in the previous technique.

In the determination of $3 \cdot 10^{-6}\%$ nickel, the coefficient of variation equals 15%.

*Determination of nickel in citric and tartaric acids.** A standard solution with an Ni^{2+} concentration of $1 \mu g/ml$ is prepared according to the technique described above.

First, a 2 g tartaric (or citric) acid sample is calcined in a crucible at 600°C to complete destruction. After cooling, 0.2 ml 6 N potassium hydroxide containing 0.1 mg/ml dimethylglyoxime is added, as well as 20 ml water. The crucible walls are carefully rinsed with this solution. After 1 hr, the solution is transferred to the electrolytic cell and the electrochemical circuit is completed with a sodium nitrate solution. Electrolysis of the stirred solution is carried out for 10 min at 0.8 V. The analysis is continued as outlined in the technique for determining nickel in ammonium molybdate and sodium tungstate.

The sensitivity of the determination is $1 \cdot 10^{-5}\%$. The sensitivity is increased to $5 \cdot 10^{-7}\%$ when the sample is increased fivefold and the concentrating step time to 1 hr.

The coefficient of variation is 15—20%.

*Determination of nickel in margarin.*** First, a 5 g margarin sample is placed in a crucible, a filter paper wick is placed in the sample and ignited. Then the residue is incinerated in a muffle furnace and dissolved in nitric acid with a density of

---

* The technique was proposed by E. Ya. Sapozhnikova.
** The technique was proposed by V. M. Vdovina and I. Kharabadzhi.

$1.4\,g/cm^3$ with hydrogen peroxide.   The solution is evaporated to dryness and the residue dissolved in doubly distilled water. The solution is transferred to a 50 ml volumetric flask, 2.5 ml in potassium hydroxide and 0.2 ml $5 \cdot 10^{-4}$ M alcoholic dimethyl-glyoxime are added, the volume is diluted to 50 ml with water, and the solution allowed to stand for 1 hr.   Then, 20 ml of the resulting solution are transferred to the electrolytic cell and the electrolysis of the stirred solution is carried out for 3 min at $+0.8$ V.

The analysis is continued as outlined in the technique for determining nickel in ammonium molybdate and sodium tungstate.

The sensitivity of the determination is $2 \cdot 10^{-5}\%$.

Chapter IV

CATHODIC STRIPPING VOLTAMMETRY OF ANIONS

In this variant, we examine polarization curves for the electrochemical reduction of an insoluble compound composed of the electrode material and anions from the solution.

FORMATION AND STRIPPING OF COMPOUNDS ON THE SURFACE OF AN ELECTROACTIVE ELECTRODE

The formation of a deposit on the surface of an electroactive electrode is possible when the electrode is anodically polarized as a result of 1) the electrochemical interaction of the electrode metal with anions adsorbed on its surface, or 2) as a result of the ionization of the metal with a subsequent chemical reaction. In the first case, we should expect the appearance of only a monolayer of compound, and in the second, growth of the deposit is possible. The experimental data indicate that salt formation on the electrode-solution interface is not usually limited to amounts corresponding to a monolayer deposit and in most cases apparently occurs according to the second type.

The ionization of a metal with a subsequent chemical reaction is described by the equations:

$$
\begin{array}{l}
\mathrm{Me} \longrightarrow \mathrm{Me}^{n+} + n e \\[4pt]
\mathrm{Me}^{n+} + n\mathrm{A}^- \longrightarrow \mathrm{MeA}_n \qquad \downarrow \\[2pt]
\hline
\mathrm{Me} + n\mathrm{A}^- \longrightarrow \mathrm{MeA}_n\!\downarrow + n e
\end{array}
\qquad\qquad (\text{IV}.1)
$$

where Me is the electrode metal and A^- is the anion which forms an insoluble compound with the Me^{n+} ions (for simplicity, the A^- is assumed univalent).

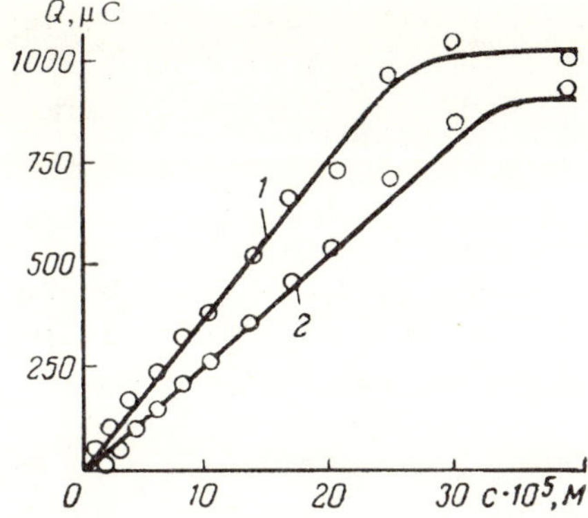

FIGURE 50. Dependence of the amount of deposited salts on the Cl^- and I^- concentrations in 0.1 N HNO_3 for $\tau_1 = 150$ sec:

1) mercurous chloride, $\varphi = -0.05$ V;
2) mercurous iodide, $\varphi_{el} = -0.3$ V.

The electrode process consists in the transport of A^- ions from the solution bulk to the electrode surface, ionization of metal atoms which make up the electrode, and the chemical reaction forming the deposit.

The deposition, i. e., the concentrating of anions, is usually carried out under adequate anodic polarization of the electrode in the presence of a supporting electrolyte (deposition under the conditions for the limiting diffusion current of the anions). The rate-limiting step is the transport of the anions to the electrode surface.

A necessary condition for the formation of a deposit on the electrode surface is a high chemical reaction rate compared to the rate for the removal of metal ions to the solution bulk, precisely as in the case discussed in Chapter III. The dependence of the amount of deposit formed in the electrolysis on the concentration of the anions participating in the reaction and the deposition time is fairly complicated and is determined by the nature of the compound formed on the electrode-solution interface, its electroconductivity, the structure of the film, and its permeability to cations of the metal being dissolved and to the anions combining with them. On the whole, this dependence is similar to that presented in the previous chapter (see Figures 29 and 30 and equations (III.9) − (III.11)). Figure 50 presents the dependence of the amount of mercurous chloride and mercurous iodide which form on the electrode in the electrolysis of 0.1 N nitric acid containing Cl^- and I^- ions in varying concentrations. Within certain limits, the amount of compound deposited is proportional to the concentration of the anions capable of reacting in the solution. Compound formation ceases when the anion concentration or the electrolysis time is large. A mercury electrode with an area of 0.02 cm^2 is made passive by an amount of deposit corresponding to $1{,}000 \mu C$ in the case of mercurous chloride and $890 \mu C$ in the case of mercury iodide. The average thickness of the passivating layer is $1.5 \cdot 10^{-5}$– $2 \cdot 10^{-5}$ cm. Thus, deposition of compounds on the electrode surface with subsequent recording of the amount of deposit formed obviously opens additional possibilities for the study of metal passivation processes.

The region with a linear dependence electrode deposit on the concentration of reactant anions and the deposition time is fairly large. The limiting value of $c\tau_1$ is approximately 0.045 g-ion-sec/liter for mercurous iodide. The range for the linear concentration dependence of the amount of deposit increases with decreasing prior electrolysis time. The segment OA (see Figure 30) is determined from the minimum anion concentration at which formation of the compound becomes possible. If the concentration of the sample anions is significantly greater than this value, the line AB arises virtually from the coordinate origin. The dependence represented in Figures 30 and 50 has a more complicated form if the sample anions are specifically adsorbed on the electrode. In this case, additional regions appear on segment OA. The possibility of such absorption-concentrating of anions will be considered below for iodide ions.

The stripping of the electrode deposit is described by an overall equation which is the opposite of relation (IV.1):

$$MeA_n + ne \longrightarrow Me + nA^- \tag{IV.2}$$

The layer adjacent to the electrode is reconstructed during the stripping step. Since the electrode reaction product is a free metal, apparently two types of reconstruction are possible depending on the radius of the anions appearing in the deposit and the initial thickness of the deposit layer.

1. The increase in the linear dimensions of the electrode from the liberated metal is much smaller than the thickness of the deposit layer stripped. If there is no mechanical removal of the outer deposit layers in the course of stripping the layers adjacent to the electrode surface, the reagent must somehow be brought to the electrode surface, and the stripping mechanism discussed in Chapter III appears possible in this case.

2. The accretion to the electrode surface during reduction from the metal liberated from the deposited compound is comparable to the thickness of the reduced part of the deposit. Then we may assume that the electroactive substance is always present on the electrode surface. The stripping kinetics is determined by the rates of discharge and ionization, the rate of removing the anions to the solution bulk, and the activity of the insoluble compound deposit on the electrode surface, which depends on the amount of this compound.

The spatial distribution of the anions during dissolution is described by the equation

$$\frac{\partial c}{\partial t} = D \frac{\partial^2 c}{\partial x^2} \tag{IV.3}$$

where c is the concentration of anions liberated during stripping of the deposit (the assumption that the rate of the chemical dissolution into the solution which is not saturated with the compound which forms the electrode deposit is slow compared to the rate of electrochemical dissolution is valid when the potential scanning rates are above 0.01 V/sec).

The boundary and initial conditions are:

$$-D \frac{\partial c}{\partial x}\bigg|_{x=0} = k_s u \exp\left[-\frac{anF}{RT}(\varphi-\varphi^0)\right] - k_s c(0,t) \exp\left[\frac{\beta nF}{RT}(\varphi-\varphi^0)\right] \tag{IV.4}$$

The first term on the right-hand side of the equation is the reduction rate of the deposit compound, and the second term is the ionization rate of the metal. The activity of the compound deposited on the electrode surface (a) is:

$$a = a_\infty \left\{1 - \exp\left[-\gamma\left(Q - \int_0^t i\,d\tau\right)\right]\right\} \tag{IV.5}$$

$\varphi = \varphi_1 - vt$ (cathodic voltage scanning)

When $x \longrightarrow \infty$ or $t = 0$, $c \longrightarrow c^0$; c^0 is the concentration of anions capable of reacting in the solution bulk.

The irreversible stripping of the compound is described by an abridged form of equation (IV.4). In this case, we may neglect the rate of the anodic process.

Then,

$$i = -nFSD\frac{\partial c}{\partial x}\bigg|_{x=0} = nFSk_s a \exp\left[-\frac{\alpha nF}{RT}(\varphi - \varphi^0)\right] \qquad (IV.6)$$

When the process is reversible (i.e., all steps except removal of the reaction products to the solution bulk occur rapidly), the electrode potential at any given moment is the equilibrium potential with respect to the activity of the compound and the concentration of potential-determining ions near the electrode surface:

$$\varphi = \varphi^0 + \frac{RT}{nF}\ln\frac{a}{cf} \qquad (IV.7)$$

where c and f are the concentration and activity coefficient of the reacting ions in the layer adjacent to the electrode.

Let the linear scanning of the potential begin from the potential φ_1, which is the equilibrium potential with respect to the initial anion concentration and the initial activity of the compound deposited on the electrode. When a rotating-disk electrode is used, the expression for the electrode potential may be written similarly to equations (II.17) − (II.20) as

$$\varphi = \varphi^0 + \frac{RT}{nF}\ln\frac{a_\infty[1 - \exp(-\gamma Q)]}{c^0 f} - vt =$$

$$= \varphi^0 + \frac{RT}{nF}\ln\frac{nFSDa_\infty\left\{1 - \exp\left[-\gamma\left(Q - \int_0^t i\,d\tau\right)\right]\right\}}{i\delta f} \qquad (IV.8)$$

Equations (IV.3) − (IV.8) are analogous to equations (II.2) − (II.5), (II.8) − (II.9), and (II.20). Mathematically, these equations differ by the values of constants. In place of the coefficients

v and $B = \dfrac{\varrho nF}{RT}$ or $B = \dfrac{nF}{RT}$ appearing in (II.8) and (II.9), the coefficients $-v$ and $B = -\dfrac{\alpha nF}{RT}$ or $B = -\dfrac{nF}{RT}$ must be used here. The method of solving the equations does not differ from that given in Appendices II and III.

The fundamental relationships between the parameters for the stripping of a microphase compound from the surface of an electroactive electrode are:

Irreversible process

$$i_{max} = \frac{anF}{RT} vQ \exp \left\{ \frac{RTSk_S a_\infty \gamma}{av} \exp \left[-\frac{anF}{RT} (\varphi_1 - \varphi^0) \right] - 1 \right\}^*$$

$$\varphi_{max} = \varphi^0 - \frac{RT}{anF} \ln \frac{av}{RTSk_S a_\infty \gamma}$$

Reversible process
rotating electrode

$$i_{max} = \frac{nF}{RT} vQ \exp \left(\frac{RT}{nFv\tau_1} - 1 \right)^*$$

$$\varphi_{max} = \varphi_p - \frac{RT}{nF} \ln \frac{nFv}{RT} \tau_1$$

fixed electrode

$$i_{max} = 0.79 \frac{nF}{RT} vQ \exp \left(\frac{\delta}{\sqrt{\frac{nF}{RT} Dv \tau_1}} - 1 \right)^*$$

$$\varphi_{max} = \varphi_p - \frac{RT}{nF} \ln \frac{\tau_1 \sqrt{\frac{nF}{RT} Dv}}{\delta}$$

These expressions show that the electrochemical dissolution of compounds from the surface of an electroactive electrode is described by a polarization curve with a current peak directly proportional to the amount of compound deposited on the electrode.

Since, under certain conditions, the amount of compound deposited on the electrode is a function only of the concentration of the anions which go into the deposit, the maximum cathodic current is a measure of the concentration of these anions in the solution. Recalling equalities (III.10) and (III.11), we may easily

---

$^*\quad \dfrac{RTSk_S a_\infty \gamma}{av} \exp \left[-\dfrac{\alpha nF}{RT} (\varphi_1 - \varphi^0) \right], \quad \dfrac{RT}{nFv\tau_1},$

$\dfrac{\delta}{\sqrt{\dfrac{nF}{RT} Dv \tau_1}} \ll 1.$ The corresponding values of the exponents are about e^{-1}.

show that the maximum stripping current is directly proportional
to the anion concentration if it is fairly high, and linearly dependent on the concentration if it is commensurate with the
minimum concentration at which formation of the compound is
possible. This behavior allows use of the deposition of insoluble
compounds on the surface of an electroactive electrode with
subsequent recording of the stripping current as a method for the
determination of microgram quantities of anions.

The dependence of the shape and position of the polarization
curve relative to the potential axis enables us to study the
kinetics of the electrochemical reduction of the deposited
compounds.

USE OF CATHODIC STRIPPING VOLTAMMETRY OF ANIONS TO STUDY THE KINETICS OF ELECTRODE PROCESSES

The nature of the dependences of the maximum current for
the stripping of a deposited compound on the quantity of deposit
on the electrode and of the potential of the maximum on the
logarithm of the potential scanning rate is determined by the
reversibility of the process. Typical calculated values are
given on page 139.

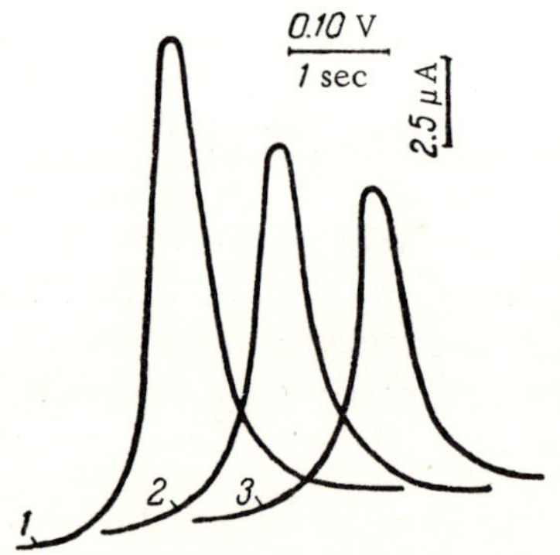

FIGURE 51. Polarization curves for
the stripping ($v = 0.1$ V/sec) of salts
deposited for $\tau_1 = 20$ sec from 0.1 N
HNO_3 solutions under varying conditions:

1) $[Cl^-] = 2 \cdot 10^{-5}$ g-ion/liter, $\varphi_{el} = -0.05$ V;
2) $[Br^-] = 2 \cdot 10^{-5}$ g-ion/liter, $\varphi_{el} = -0.15$ V;
3) $[I^-] = 3 \cdot 10^{-5}$ g-ion/liter, $\varphi_{el} = -0.3$ V.

Points of inflection are observed at $i_1 = 0.72\, i_{max}$ and $i_2 =
= 0.53\, i_{max}$. The calculations are given for $T = 298°C$.

Fundamental parameters of the process	Reversible process		Irreversible process
	rotating electrode	fixed electrode	
$\tan\theta = \dfrac{di_{max}}{dQ}$	$14.2\ nv$	$11.2\ nv$	$14.2\ \alpha nv$
$\tan\zeta = \dfrac{d\varphi_{max}}{d\log v}$	$\dfrac{0.059}{n}$	$\dfrac{0.030}{n}$	$\dfrac{0.059}{\alpha n}$
b	$\varphi_e - \dfrac{0.059}{n}\log 38.8\,n\tau_1$	$\varphi_e - \dfrac{0.059}{n}\log\dfrac{6.2\tau_1\sqrt{nD}}{\delta}$	$\varphi^0 - \dfrac{0.059}{\alpha n}\log\dfrac{\alpha}{RTSk_S a_\infty \gamma}$
$\Delta\varphi_n$	—	—	$\dfrac{1.87}{\alpha}$
$i = f(t)$ when φ=constant	—	Hyperbolic	Has a section where i = const.

where $\tan\theta$ and $\tan\zeta$ are the slopes of the linear plots describing $i_{max}=f(Q)$ and $vt_{max}=f(\log v)$; b the intersection by the line $vt_{max}=f(\log v)$ with the ordinate axis, and $\Delta\varphi_n$ the potential difference between the points of inflection of the ascending and descending branches of the polarization curve.

Criteria for the reversibility of the processes are found in the slopes of the linear plots describing the functions $i_{max} = f(Q)$ and $f_{max} = f(\log v)$ and the potential difference between the points of inflection on the polarization curves for the stripping of microphases from the surface of an electroactive electrode. The nature of the change in the current for the stripping of a fairly large amount of deposit from the surface of a fixed electrode at constant potential serves as an additional test.

Typical polarization curves for the cathodic stripping of mercurous chloride, mercurous bromide, and mercurous iodide are presented in Figure 51. The maximum stripping current is directly proportional to the amount of the respective compound deposited on the electrode surface.

The maximum current potential for the stripping of mercurous iodide, as evidenced in Figure 52, is linearly dependent on the electrode potential scanning rate. The time-dependence of the stripping current of mercury (I) chloride at a constant electrode potential recorded on an OP-1 oscillographic polarograph is presented in Figure 53. There is a sharp difference in the nature of the dissolution of mercury (I) chloride in solutions which have been subjected and those not subjected to filtration through charcoal. The corresponding $i—t$ plots for mercurous bromide and mercurous iodide differ somewhat less.

Functions characterizing the stripping kinetics of a mercury (I) chloride microphase are:

Functions	Slopes	
	water purified*	water not purified
$i_{max} = f(Q)$ $v = 0.1$ V/sec, $n = 1$	1.1	0.91
$\varphi_{max} = f(\log v)$; $n = 1$	0.03	0.09
$i = f(t)$ at $\varphi = $ const., $v = 0$; $n = 1$ ...	Hyperbolic	With a section in which i is constant

* Purifying of water from surface-active substances by filtering through charcoal.

Comparing these results with the calculated data (p. 139), we easily conclude that when a solution which has been filtered through charcoal is used, the process is reversible, while if the solution has not been purified, the process is irreversible, i. e., the slow step is electron transfer. The value of the coefficient α calculated from the experimental data is 0.64.

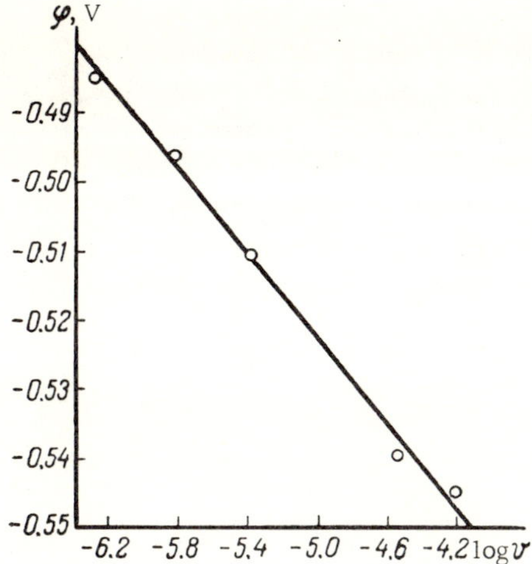

FIGURE 52. Dependence of the maximum current potential for the stripping of mercurous iodide deposited from a solution 0.1N in KNO_3 and $3 \cdot 10^{-5}$ M in I^- for $\tau_1 = 60$ sec and $\varphi_{el} = -0.3$ V on the logarithm of the potential scanning rate. (The supporting solution was filtered through charcoal.)

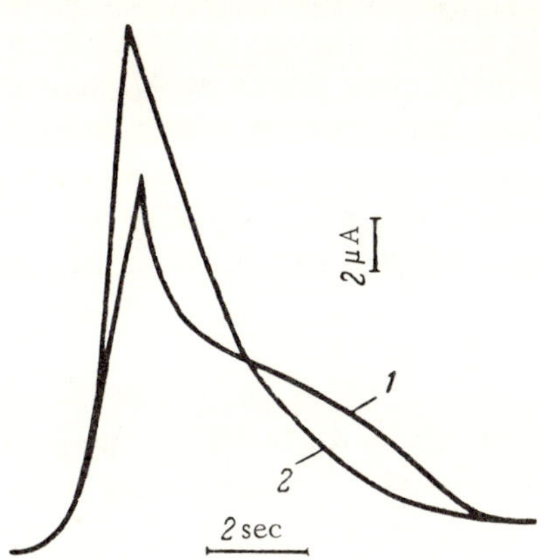

FIGURE 53. Dependence of the stripping current of mercury (I) chloride deposited from a solution 0.1 N in HNO_3, and $6 \cdot 10^{-5}$ M in Cl^- for $\tau_1 = 30$ sec and $\varphi_{el} = -0.05$ V on the time at a constant electrode potential of -0.27 V:

1) supporting solution not filtered through charcoal; 2) supporting solution filtered through charcoal.

The results agree with the data of Cornish et al. /144/, who used a galvanostatic method to show that the exchange current for the formation of mercury (I) chloride decreases by a factor of 20—30 as a result of the adsorption of traces of surface-active substances usually present in solutions on the electrode.

USE OF CATHODIC STRIPPING VOLTAMETRY IN THE ANALYSIS OF ANIONS

Anions are concentrated in the form of an insoluble compound with metal ions from the electrode, which forms in the anodic polarization process. For this purpose, a standard mercury capillary electrode is used and the deposition is carried out from a stirred solution at a potential providing a steady state. This potential is usually somewhat more positive than the equilibrium potential of the corresponding electrode-insoluble compound—anion system in the solution; the anions in question are those that go into the deposited compound.

Information on the concentration of the ions forming the insoluble compound on the electrode surface is obtained by recording the polarization curves for the electrochemical reduction of the deposits and measuring the maximum cathodic current.

Determination of chloride, bromide, iodide, sulfide, and sulfate ions

Halide ions /145−149/ are deposited on the surface of a mercury electrode from potassium nitrate solutions containing not less than 10^{-6} g-ion/liter Cl^-, Br^-, or I^- at potentials of 0.35, 0.35, and 0.1 V,* respectively. Mercurous sulfide is deposited at -0.5 V from 1 N NaOH containing not less than $5 \cdot 10^{-8}$ g-ion/liter S^{2-} ions. Figures 54 and 55 present typical polarization curves for the stripping of mercurous chloride, bromide, iodide, and sulfide and the dependence of the maximum cathodic currents on the concentration of the corresponding anions. In all cases with the exception of mercurous iodide, the maximum current for the stripping of the deposited compound is directly proportional to the concentration of the corresponding anions. Obviously, the minimum determinable concentrations of Cl^-, Br^-, and S^{2-} ions are below the concentrations cited in Figure 55. The linear dependence of the maximum stripping current of mercurous iodide on the I^- ion concentration, as will be shown below, is due to a different mechanism of concentrating these anions. The shape of the polarization curves for the electrochemical dissolution of mercurous halide salts remains unchanged when the supporting electrolyte concentration is altered and the pH of the solution varied from 0 to 5. At higher

---

* The potential values are given relative to the saturated calomel electrode. A mercurous sulfate electrode is used in the measurements in order to avoid contaminating the solution under analysis with chloride ions. The potential of the mercurous sulfate electrode containing a saturated K_2SO_4 solution acidified to pH 1 with sulfuric acid equals 0.390 ± 0.002 V relative to the saturated calomel electrode at 18°C. Where the reference electrode is not specifically indicated, the potentials are given relative to the saturated calomel electrode.

pH, mercurous hydroxide* forms on the electrode-solution interface. This interferes with the determination of halide ions,**
since the mercurous hydroxide reduction current masks the
reduction current for the mercurous halide salts.

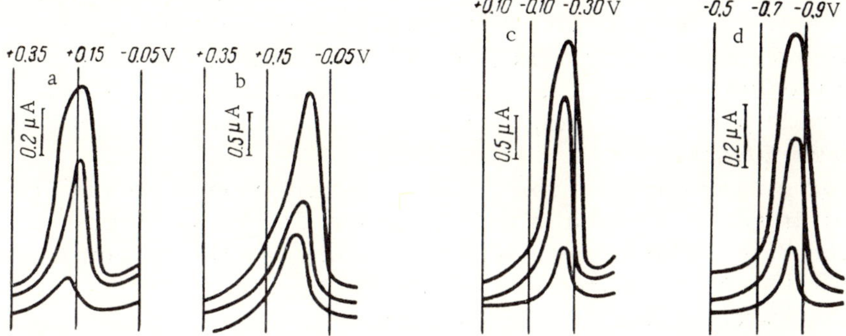

FIGURE 54. Polarization curves for the stripping of mercurous chloride (a), bromide
(b), and iodide (c) deposited from a solution 0.05 N in KNO_3, and mercurous sulfide (d)
deposited from a 1 N NaOH solution.

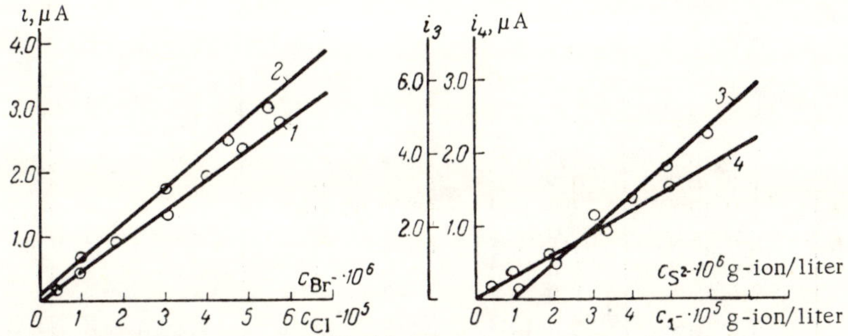

FIGURE 55. Dependence of the maximum current for the stripping of
mercurous chloride (1), bromide (2), iodide (3), and sulfide (4) on the concentration of Cl^-, Br^-, I^-, and S^{2-}, respectively.

---

* The current for the electrochemical reduction of mercurous hydroxide is very sensitive
to a change in the OH^- ion concentration and may be used for a very fine measure of
the pH between 6 and 7. The optimal potential for the concentrating step is 0.40 V.
The maximum cathodic current is observed in the 0.1—0 V region.

** It will be shown below that mercurous hydroxide can be removed from the electrode
surface chemically and then the polarogram for the electrochemical reduction of mercurous chloride and bromide recorded.

The polarization curves for the reduction of mercury (I) chloride are strongly distorted as the result of adsorption of surface-active substances which are always present in solutions not subjected to special purification. The polarograms of mercurous iodide and sulfide are less sensitive to these contaminations. The effect of surface-active substances is eliminated if the electrode is cathodically polarized under a current of hydrogen in 1 N nitric acid (a current of 30 mA) for 1 min after each measurement. This makes it possible to use a mercury electrode prepared a single time for several consecutive determinations.

The mean square error in the determination of $5 \cdot 10^{-6}$ g-ion/liter Cl^- is 7–8% and the maximum error (3σ) does not exceed 23%. The same results are found for the determination of Br^-, I^-, and S^{2-} concentrations. The accuracy of the analysis is illustrated by the data in the table.

Determination of Cl^-, I^-, S^{2-} ions by different methods

Ion	Concentration, M	Concentration (M) found by the methods			
		potentiometric	amperometric	mercurimetric	cathodic stripping voltammetry
Cl^-	$5.0 \cdot 10^{-5}$	$5.3 \cdot 10^{-5}$	$4.6 \cdot 10^{-5}$		$5.2 \cdot 10^{-5}$
	$1.0 \cdot 10^{-4}$	$0.98 \cdot 10^{-4}$	$1.05 \cdot 10^{-4}$		$1.1 \cdot 10^{-4}$
	$2.0 \cdot 10^{-4}$	$2.0 \cdot 10^{-4}$	$2.05 \cdot 10^{-4}$		$1.9 \cdot 10^{-4}$
I^-	$5.0 \cdot 10^{-5}$	$5.2 \cdot 10^{-5}$	$5.0 \cdot 10^{-5}$		$5.3 \cdot 10^{-5}$
	$1.0 \cdot 10^{-4}$	$1.0 \cdot 10^{-4}$	$0.95 \cdot 10^{-4}$		$1.05 \cdot 10^{-4}$
	$2.0 \cdot 10^{-4}$	$1.9 \cdot 10^{-4}$	$2.1 \cdot 10^{-4}$		$1.95 \cdot 10^{-4}$
S^{2-}	$5.2 \cdot 10^{-5}$	$5.3 \cdot 10^{-5}$	$5.2 \cdot 10^{-5}$	$5.2 \cdot 10^{-5}$	$5.0 \cdot 10^{-5}$
	$1.1 \cdot 10^{-4}$	$1.1 \cdot 10^{-4}$	$1.0 \cdot 10^{-4}$	$1.1 \cdot 10^{-4}$	$1.2 \cdot 10^{-4}$
	$2.3 \cdot 10^{-4}$	$2.2 \cdot 10^{-4}$	$2.4 \cdot 10^{-4}$	$2.3 \cdot 10^{-4}$	$2.1 \cdot 10^{-4}$

The minimum determinable* concentration of Cl^- at 25°C is $5 \cdot 10^{-6}$ g-ion/liter and of Br^-, $1 \cdot 10^{-6}$ g-ion/liter (the deposition time is 10 to 20 min). The sensitivity of the determinations increases when the temperature of the solution is increased or ethyl alcohol is added, i. e., by reducing the solubility of the compounds formed on the electrode. The minimum determinable

* The minimum concentration which can be determined with an error not exceeding 20%.

concentration of Cl^- ions at 2°C equals $1 \cdot 10^{-6}$ g-ion/liter in a solution 0.1 N in HNO_3 and 84% in C_2H_5OH, and at 2–3°C, the sensitivity of the determinations reaches $1 \cdot 10^{-7}$ g-ion/liter. The linear plot describing the dependence of the maximum current for the stripping of mercurous chloride on the Cl^- ion concentration for a Cl^- ion concentration of $1 \cdot 10^{-7}$ g-ion/liter arises from the coordinate origin.

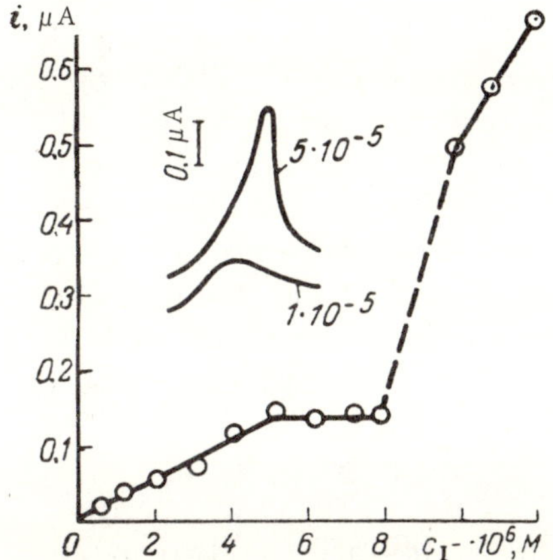

FIGURE 56. Dependence of the maximum cathodic current on the I⁻ ion concentration and polarization curves for the stripping ($v = 0.011$ V/sec) of mercurous iodide deposited for $\tau_1 = 1$ min and $\varphi_{el} = -0.3$ V from 0.1N $NaNO_3$ solutions with different I⁻ concentrations (numbers on the curves: g-ion/liter).

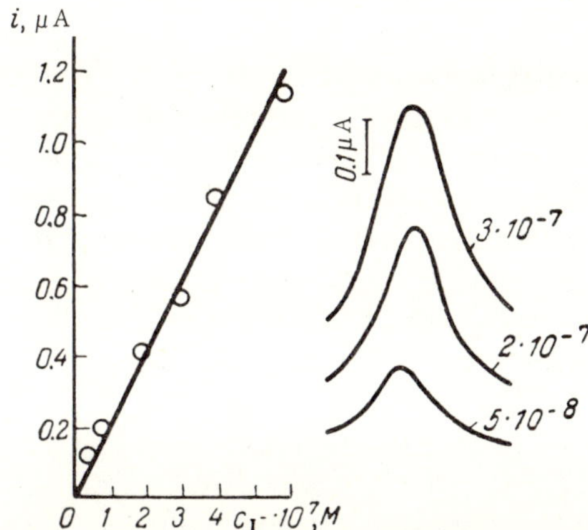

FIGURE 57. Dependence of the maximum cathodic current on the I⁻ ion concentration and polarization curves for the stripping ($v = 0.5$ V/sec) of mercurous iodide deposited for $\tau_1 = 1$ min, $\varphi_{el} = -0.3$ V from 0.1 N $NaNO_3$ solutions with different I⁻ concentrations (numbers on the curves: g-ion/liter).

The mechanism for the concentrating iodide ions is probably different in different concentration ranges (Figures 56 and 57). Iodide ions are perhaps concentrated from extremely dilute solutions as a result of specific adsorption but crystals of mercurous iodide appear on the electrode surface at higher I⁻ concentrations. To determine extremely low I⁻ ion concentrations (region of adsorption concentrating) at which the observed net signal has a low absolute value, it is wise to carry out the cathodic polarization of the electrode at a high potential scanning rate to increase the signal using an oscillographic polarograph. According to Figure 57, the minimum determinable I^- ion concentration in this case reaches $5 \cdot 10^{-8}$ g-ion/liter. The oscillographic recording of polarization curves is also expedient in the determination of S^{2-} ions at concentrations less than $5 \cdot 10^{-7}$ M giving increased sensitivity ($5 \cdot 10^{-8}$ g-ion/liter). In other anion determinations, use of an oscillographic polarograph does not increase the sensitivity of the conventional analysis, since the sensitivity is limited by the solubility product of the deposited compounds.

The electrochemical formation and dissolution of mercurous sulfide is used to determine SO_4^{2-} ions, since the latter cannot be concentrated directly owing to the high solubility of sulfates. Sulfate ions are first reduced to S^{2-} and then concentrated on a mercury electrode /150/. The reduction is carried out using a titanium (III) phosphate solution. The hydrogen sulfide formed is absorbed by an alkali solution /151/ and the sulfide ion concentration is determined. The experimental technique and determination results are given below for analyzing aluminum nitrate.

Techniques have been developed for the determination of traces of chloride ion in potassium nitrate /145/, aluminum nitrate /152/, uranyl sulfate /148/, the tungstates and molybdates of potassium and strontium /153/, and sodium acetate /154/, of iodide ions in barium carbonate, strontium carbonate, calcium carbonate /155/, sodium nitrate /147/, and sodium acetate /154/, of sulfide ions in water /156/, and of sulfate ions in aluminum nitrate and alkali metal carbonates /150/.

*Determination of chloride ions in
aluminum nitrate /152/*

REAGENTS AND APPARATUS

Distilled water is prepared in an apparatus without rubber parts by distillation of a solution containing 50 ml H_2O, 2 ml HNO_3, and 0.5 g $AgNO_3$.

Nitric acid. An 0.5 N solution distilled in a quartz apparatus from a mixture of 500 ml HNO_3 and 0.5 g $AgNO_3$; an 0.1 N solution (for polarization of the mercury electrode).

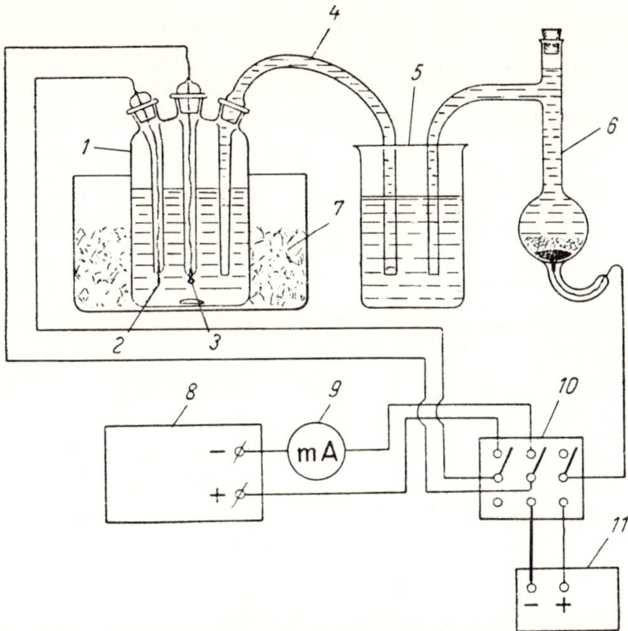

FIGURE 58. Apparatus for the determination of chlorides:

1) electrolytic cell; 2) a platinum igniting electrode ($d = 0.5$ mm, $l = 5$–10 mm); 3) mercury indicator electrode; 4) salt bridge; 5) intermediate vessel; 6) mercurous sulfate electrode; 7) ice bath; 8) rectifier model VSA-5; 9) milliammeter; 10) switch; 11) polarograph model ON-101.

Ethyl alcohol distilled from a mixture of 500 ml C_2H_5OH and 0.5 g $AgNO_3$.

Solution for preparing the mercury electrode: 10 g $Hg_2(NO_3)_2$, 40 ml H_2O, and 0.05 ml concentrated nitric acid.

Apparatus for determination of chlorides (Figure 58).

Procedure. First, a 5 g moist sample is placed in a graduated cylinder and dissolved in 4 ml 0.5 N nitric acid and 21 ml distilled alcohol. The solution is transferred to the electrolytic cell, which is placed in an ice bath. The measurements are begun after the temperature of the solution is reduced to 2–3°C. The salt bridge and intermediate vessel are filled with 0.5 N nitric acid. The mercury electrode is polarized cathodically (see Chapter V), the negative pole of the polarograph is

connected to the mercury electrode, and the positive pole to the
mercurous sulfate electrode.   The stirrer is turned on and the
electrolysis is carried out for ·5 min at a potential of 0 V.
Chloride ions are concentrated on the electrode as mercury (I)
chloride.   The stirring is stopped, the solution is allowed to
settle for 10–15 sec, and the cathodic polarization curve for the
stripping of mercurous chloride is recorded from 0 to -1.0 V.
The maximum cathodic current, which is observed at a potential
from -0.2 to -0.4 V, is measured.   The concentration of Cl^- ions
is found from a known addition.   For this purpose, 0.5 ml stan–
dard chloride solution is introduced into the cell.   Electrodes
2 and 3 are connected to the rectifier (the mercury electrode to
the minus), and the polarization is carried out for 30 to 40 sec
with a current of 5 mA.   Then, the mercury electrode is connected to the polarograph and the analysis is repeated.   In the
determination of $5 \cdot 10^{-6}\%$ chloride, the coefficient of variation
is 20%.

Determination of chloride ions in the tungstates and molybdates of potassium and strontium /153/

REAGENTS AND APPARATUS

Distilled water previously boiled with an alkaline solution
of potassium permanganate for 4 to 5 hr and then doubly distilled
in an apparatus not containing rubber parts.
   Zinc amalgam.   Metallic zinc (1.5 g) washed with 1 N nitric
acid is dissolved in 50 ml mercury.   The amalgam is kept under
water and renewed after every five determinations in the analysis
of tungstates and after every two determinations in the analsis
of molybdates by adding 1.5 g zinc.
   Procedure.   First, a 1 g sample is dissolved in 20 ml 25%
tartaric acid with heating.   Several drops of concentrated nitric
acid are added and the solution is transferred to a separating
funnel.   The zinc amalgam is added, and the mixture is shaken
until the solution becomes colorless and transparent in the
analysis of tungstates or transparent green in the analysis of
molybdates.   Then, the amalgam is removed and the solution is
transferred to the electrolytic cell.   The salt bridge is filled

with an 0.1 N potassium sulfate solution.   The mercury electrode
is polarized cathodically by connecting it to the negative pole of
the polarograph and the mercurous sulfate electrode to the
positive, and the electrolysis of the stirred solution is carried
out for 15 min at a mercury electrode potential of -0.05 V.   The
chloride ions are concentrated on the mercury electrode as
mercury (I) chloride.   The stirring is stopped, and after 15–
20 sec the cathodic polarogram for the stripping of mercurous
chloride is recorded.   The maximum current is observed at
from -0.25 to -0.35 V.   The chloride ion concentration is found
from a calibration graph or from a known addition.   An addition
of 0.005 mg-ion of Cl^- corresponds to a concentration of
$5 \cdot 10^{-4}\%$.   In the determination of $5 \cdot 10^{-4} - 1 \cdot 10^{-2}\%$ chloride, the
coefficient of variation does not exceed 15%.

*Determination of iodide ions in barium carbonate, strontium
carbonate, and calcium carbonate /155/.*   First, a 2 g carbonate
sample is placed in a graduated cylinder and dissolved in 15 ml
water with the dropwise addition of 2 ml concentrated nitric acid.
After the gas evolution ceases, the solution volume is brought up
to 20 ml and then transferred to the electrolytic cell.   The salt
bridge and the intermediate vessel are filled with 0.1 N potassium sulfate, and oxygen is removed by bubbling nitrogen for
10 min.   The mercury electrode is connected to the negative pole
of the polarograph, and the mercurous sulfate electrode to the
positive pole.   The electrolysis of the stirred solution is carried
out for 1 min at a mercury electrode potential of -0.3 V.   Iodide
ions are concentrated on the electrode as mercury (I) iodide.
The stirring is stopped and after the solution settles (20–30 sec),
the cathodic oscillogram is recorded at a voltage scanning rate
of 0.5 V/sec at from -0.3 to -0.8 V.   The maximum cathodic
current is observed at from -0.45 to -0.55 V.   The iodide ion
concentration is found from a calibration graph.

In the determination of $5 \cdot 10^{-6} - 5 \cdot 10^{-5}\%$ iodide, the coefficient of variation is 15%.

*Determination of sulfate ions in aluminum nitrate**

REAGENTS AND APPARATUS

Reagent for the reduction of sulfates.   First, 1.5 g titanium
is dissolved in 50 ml orthophosphoric acid with heating.   Then

---

* The technique was proposed by E.Ya.Sapozhnikova.

20 ml of the resulting solution are placed in the apparatus shown in Figure 59. Air is expelled from the apparatus with carbon dioxide introduced through a tube reaching down to the bottom of the reaction flask. Then the fitting is changed, and after a continuous current of carbon dioxide is established, the solution is boiled without interrupting the gas flow for 30 to 60 min. The sulfate impurity in the reagents is reduced and the hydrogen sulfide formed is removed from the reaction mixture. The resulting reagent solution may be used for seven to ten analyses. Air must not be allowed to come into contact with the solution. After each determination, 5 ml of water should be added.

Acid-trapping solution, 5—7 N in KOH and 0.5 N in $N_2H_4 \cdot HCl$.

Distilled water, barium nitrate is added to the water before distillation.

Apparatus for reduction of sulfate ions.

A hanging mercury drop and saturated calomel electrodes are employed.

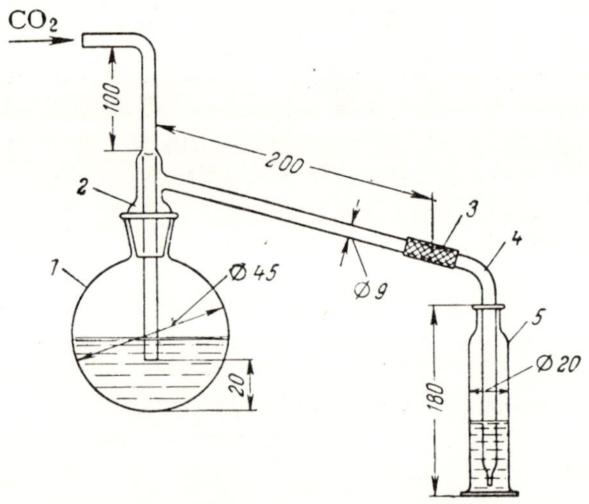

FIGURE 59. Apparatus for reduction of sulfate ions:

1) quartz reaction flask; 2) quartz fitting; 3) polyethylene sleeve; 4) glass tube; 5) glass trap.

Procedure. First a 15—20 g sample is roasted at 300°C until the nitrates decompose (about 1.5 hr). Then an 0.8 g calcined sample is placed in a cooled reaction flask (see Figure 59), the walls are washed with 5 ml water, and carbon dioxide is bubbled through the solution for 15 min. The trap is filled with the KOH-$N_2H_4 \cdot$HCl solution for absorbing S^{2-}, and the reducing solution is carefully heated in flask 1 to boiling while carbon dioxide is passed at a rate of two to three bubbles per second for 40 min from the moment the reaction mixture begins to boil. Then the solution volume in the trap is brought up to 20 ml with distilled water, and this solution is transferred to the electrolytic cell. The electrolysis of the stirred solution is carried out for 1 min at -0.5 V and the polarization curve for the stripping of mercurous sulfide is recorded. The maximum stripping current, which is observed on the polarogram at from -7.5 to -1.0 V, is measured.

The sulfate concentration in the sample is found from a known addition by introducing 0.5 ml of a standard potassium sulfate solution into the cooled reaction mixture, washing the walls of the vessel with 5 ml of water, and repeating the analysis. Before the polarogram is recorded, the electrode is polarized at -1.2 V for 40 sec.

Concentrations, ranging from $5 \cdot 10^{-8}$ to $1 \cdot 10^{-5}$ g-ion/liter can be determined. Larger amounts of S^{2-} are bound in insoluble sulfides by heavy metal impurities in the trap solution. The mean relative square error in the determination of 0.005–0.01 μg/ml SO_4^{2-} is 15%.

In the analysis of nonreducible compounds such as alkali metal carbonates, the decomposition step is omitted /150/.

Determination of tungstate, molybdate, vanadate, and chromate ions /146/

Tungstate, molybdate, and vanadate ions are deposited from an 0.05 N potassium nitrate solution (pH = 5.0–5.5) on the surface of a mercury electrode at a potential of 0.40 V relative to the saturated calomel electrode. For more acidic solutions, these anions are not concentrated as readily, and in alkaline solutions mercurous hydroxide forms. The optimum pH should be obtained by introducing acids or bases. Unfortunately, buffer solutions cannot be used, since the components of buffer solutions interact with the electrode material at potentials more negative than the potentials for deposition of the mercury salts of the elements being determined, are concentrated on the electrode, and interfere with the determination.

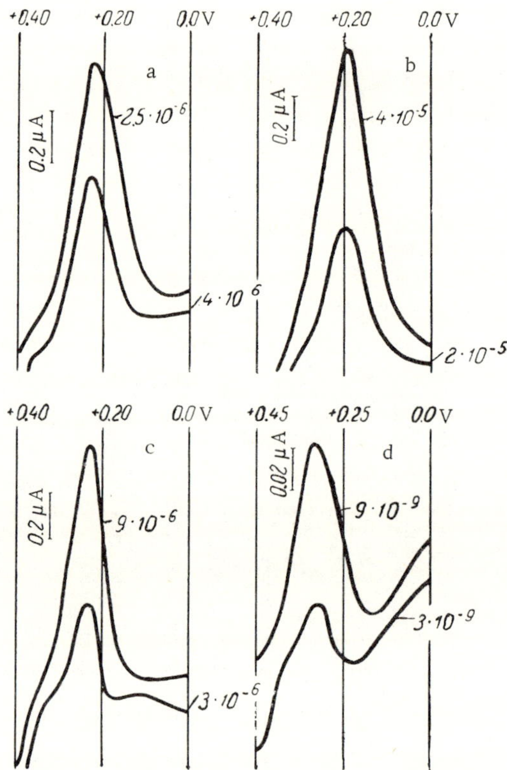

FIGURE 60. Polarization curves for the cathodic stripping ($v = 0.11$ V/sec) of mercurous tungstate (a), molybdate (b), vanadate (c), and chromate (d) deposited from 0.1 N $NaNO_3$ solutions containing various amounts of the respective ions (numbers on the curves: g-ion/liter) under different conditions:

a) and c) $\tau_1 = 5$ min, $\varphi_{el} = 0.40$ V; b) $\tau_1 = 10$ min, $\varphi_{el} = 0.40$ V; d) $\tau_1 = 10$ min, $\varphi_{el} = 0.45$ V.

For a deposition time of 10 to 20 min, the minimum determinable concentrations (in g-ion/liter) are: $4 \cdot 10^{-7}$ for WO_4^{2-} and $1 \cdot 10^{-6}$ for MoO_4^{2-} and VO_3^-. The reproducibility of the results is the same as that cited above for halide ions.

Mercurous chromate forms at + 0.45 V from an 0.05 N potassium nitrate solution (pH = 2—4). The high sensitivity of the determination ($3 \cdot 10^{-9}$ g-ion/liter for 10 min electrolysis time) is probably due to the ability of mercurous chromate to form dense films which adhere well to the electrode surface. This makes it possible to concentrate chromate ions under the conditions for mercury stripping. The high concentration of mercury ions in the layer adjacent to the electrode allows deposition of the compound at a low CrO_4^{2-} ion concentration. The electrochemical deposition of mercurous vanadate, tungstate, and molybdate may also occur under the conditions of a negligible mercury stripping current. Increasing this current causes the precipitates to stop forming on the electrode.

Determination of microtraces of chromium (VI) in cadmium sulfate /157/. A hanging mercury drop capillary and a mercurous sulfate electrode are employed.

First, a 2 g cadmium sulfate sample is dissolved in 18 ml water, the pH is brought to between 4 and 4.5 by adding 0.01 N nitric acid, the solution is diluted to 20 ml with water and transferred to the electrolytic cell. Electrolysis of the stirred solution is carried out for 10 min at a potential of 0 V. Chromium is concentrated on the electrode as mercury (I) chromate. The stirring is stopped, and the cathodic polarization curve is recorded at from 0 to -0.5 V.

The CrO_4^{2-} ion concentration is found from the maximum current for the electrochemical reduction of mercurous chromate according to a calibration graph.

The determination of chromium (VI) may be carried out in the $2 \cdot 10^{-7} - 2 \cdot 10^{-6}$% range. For higher concentrations, the electrolysis time should be decreased to 1 min. The coefficient of variation is 15%.

DETERMINATION OF SEVERAL ORGANIC ANIONS AND SUBSTANCES WITH CHARACTERISTIC FUNCTIONAL GROUPS

Organic anions and substances with characteristic functional groups capable of forming insoluble compounds with mercury are concentrated electrochemically on the mercury electrode surface in a manner similar to that of inorganic anions. The polarization curves for the electrochemical reduction of mercurous oxalate, succinate, dithizonate, and diethyldithiophosphate are presented in Figure 61. The compounds deposit on the electrode at 0.35, 0.40, 0.10, and 0 V, respectively. The best supporting electrolyte is 0.1 M potassium nitrate. To increase the solubility of dithizone, 10% acetone is introduced. Varying the pH between 2 and 4 does not affect the polarization curves.

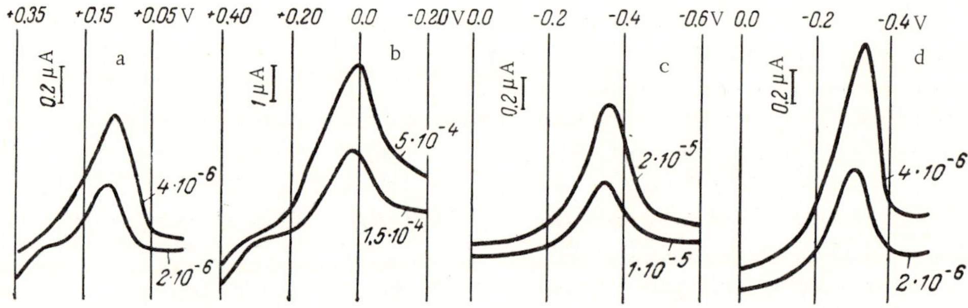

FIGURE 61. Polarization curves for the cathodic stripping ($v = 0.011$ V/sec) of mercurous oxalate (a), succinate (b), dithizonate (c), and diethyldithiophosphate (d) deposited from 1 N NaNO$_3$ solutions containing various amounts of the respective ions (numbers on the curves: g-ion/liter) under different conditions:

a) $\tau_1 = 1$ min, $\varphi_{el} = 0.35$ V; b) $\tau_1 = 1$ min, $\varphi_{el} = 0.40$ V; c) $\tau_1 = 1$ min, $\varphi_{el} = 0.10$ V; d) $\tau_1 = 5$ min, $\varphi_{el} = 0$ V.

Berge and Jeroschewski /158/ studied the behavior of several sulfur-containing organic compounds on the hanging mercury drop electrode and showed that the method described here may be used to determine small amounts of thiourea ($1 \cdot 10^{-8}$ M), 2-mercaptobenzthiazole ($1 \cdot 10^{-6}$ M), thioanalide, and rubean hydride. The maximum current for the electrochemical reduction is proportional to the concentration of the organic compounds.

INTERACTION OF ANIONS AND POSSIBILITIES OF ANALYZING MIXTURES

The possibilities of determining anions which form less soluble compounds with the electrode material in the presence

of anions whose compounds are more soluble and vice versa
will be considered separately.

First case. Foreign anions in the solution will not influence the results of the determination of the anions of interest
if the electrode potential necessary for the electrodeposition of
a film of the compound MeA_{det} is more negative than the potential
for initiating the formation of the compound MeA_{for} (here Me is
the metal of which the electrode is composed, A_{det}^- is the anion
being determined, and A_{for} is a foreign anion in the solution).

The formation of a compound on the electrode surface is
possible at a potential slightly more positive than the equilibrium
potential of the corresponding system. For a tentative calculation of the maximum permissible concentration of foreign ions
in the solution, we shall assume that the potential for concentrating the anions being determined must not be higher than the
equilibrium potential of the system

$$Me + A_{for}^- \; \rightleftharpoons \; MeA_{for} \downarrow + e \qquad (IV.9)$$

Then,

$$\varphi_{for}^0 \; - \frac{RT}{nF} \ln [A_{for}] \geqslant \varphi_{el}$$

$$\log [A_{for}] \leqslant \frac{n(\varphi_{for}^0 - \varphi_{el})}{0.059}$$

where φ_{el} is the empirical optimum potential for depositing
the ions being concentrated.

Thus, the determination of sulfide ions is not disturbed by
nearly all other anions. Chloride, bromide, and iodide may be
determined in the salts of nitric, sulfuric, hydrochloric, hydro-
fluoric, phosphoric, boric, and many organic acids. The determination of iodide is possible in the presence of a five-thousand-
fold molar excess of chloride ions or a fiftyfold excess of
bromide ions, and the determination of bromide is not disturbed
by a thirtyfold excess of chloride ions.

Second case. The determination of anions forming more
soluble compounds with the electrode material in the presence
of anions whose corresponding compounds are less soluble is
considerably more complicated. Here, a film of MeA_{for} forms

along with the deposition of a film of the salt MeA$_{det}$. The appearance of separate current peaks on the polarization curve for stripping the deposit allows determination of the anions quantitatively if the films of MeA$_{det}$ and MeA$_{for}$ form in the concentration step in amounts proportional to the concentrations of the anions A$^-_{det}$ and A$^-_{for}$. This fact and thus the possibility of determining the anions in this case have been established experimentally.

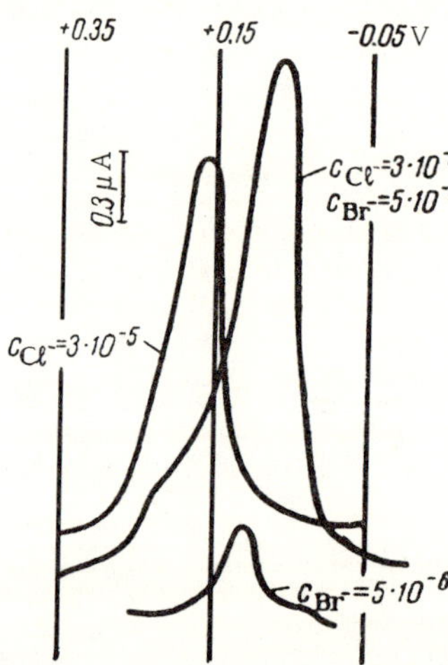

FIGURE 62. Polarization curves for the stripping ($v = 0.011$ V/sec) of salts deposited from 0.1 N KNO$_3$ solutions with various Cl$^-$ and Br$^-$ concentrations (numbers on the curves: g-ion/liters).

Typical polarization curves for the stripping of a deposit of salts formed on the electrode in the electrolysis of a solution containing chloride and bromide ions are presented in Table 62. The current maxima on the curves are in the potential range characteristic for the stripping of mercurous bromide even when the [Br$^-$]:[Cl$^-$] ratio is 1:10. The maximum current for the stripping of a deposit formed from the anodic polarization of mercury in a solution concentration and Br$^-$ is directly proportional to the total concentration of these ions. A mixture of Cl$^-$ and Br$^-$ may be analyzed using this dependence by determining the total concentration ($\varphi_{el} = 0.35$ V) and the bromide ion concentration separately ($\varphi_{el} = 0.25$ V). A quantitative determination of Cl$^-$ and Br$^-$ ions in the presence of more than $1 \cdot 10^{-5}$ g-ion/liter I$^-$ is impossible. With increasing iodide concentration, the maximum current for the stripping of mercurous chloride decreases, and the maximum current for the reduction of mercurous iodide increases. The fact that a similar dependence is also observed in the analysis of solutions containing I^- and Br^- is apparently a consequence of the predominant formation of the less soluble compound (HgI) in the concentrating step.

To determine chloride and bromide ions in the presence of iodide ions, the electrolysis of the sample solution (pH = 5–6) may be carried out at 0.6 V (the voltage on the cell is about 2 V). The iodide ions evidently are oxidized according to the reaction $2I^- \rightarrow I_2 + 2e$ rather than being concentrated and mercurous chloride, bromide, and hydroxide are deposited on the electrode. To remove the mercurous oxides whose reduction current masks the useful signal, the working electrode should be withdrawn from the solution after the concentrating step, washed with 0.1 N nitric acid and water, and voltage applied to the cell decreased to the value corresponding to the starting potential for the stripping of mercurous chloride, and then the polarization curve recorded. Applying a high anodic polarization to the electrode with the subsequent chemical removal of the oxide film allows determination of $5 \cdot 10^{-6}$ to $1 \cdot 10^{-5}$ g-ion/liter Cl^- (or Br^-) in the presence of $5 \cdot 10^{-5}$ g-ion/liter I^-. The determination of large amounts of Cl^- or Br^- ions is not disturbed by a fivefold molar excess of iodide ions.

Thus, the analysis of a solution containing Cl^-, Br^-, and I^- ions should be done in two or three steps: 1) electrolysis at a potential of + 0.10 V, determination of I^-, 2) for an I^- concentration not exceeding $1 \cdot 10^{-5}$ g-ion/liter, the determination of Br^- (deposition at 0.25 V), 3) total Cl^- and Br^- (deposition at 0.35 V). If the iodide ion concentration is greater than $1 \cdot 10^{-5}$ G–ion/liter, the total concentration of Cl^- and Br^- may be determined by the method described above.

Mercurous tungstate, molybdate, and vanadate form on the electrode under the same conditions. Regardless of the ratio of anions in solution, the cathodic polarograms have one current maximum whose value is the sum of the stripping currents of the corresponding mercurous salts.

To determine chromate ions in the presence of a commensurate amount of tungstate, molybdates, and vanadates, the pH of the solution under analysis must be lowered to 3.

The concentration of halide ions in a solution of tungstates and molybdates may be determined by reducing the tungstate and molybdate ions (see Technique for determining chloride in the tungstates and molybdates of calcium and strontium).

Chapter V

EXPERIMENTAL TECHNIQUES

ELECTRODES FOR ANODIC AND CATHODIC
STRIPPING VOLTAMMETRY OF SOLIDS

The deposition and stripping of metals and insoluble compounds are performed mainly on solid platinum /57, 69, 95, 159, 160/, gold /161–163/, and graphite (carbon) electrodes. The concentrating of iron /108/ as $Fe(OH)_3$ and rhenium /164/ as ReO_2 on the hanging mercury drop electrode has also been described. The development of anodic and cathodic stripping voltammetry has involved the use of different types of carbon electrodes. This is due to the requirements of inertness of the electrode material, the fairly high overvoltage for hydrogen and oxygen on these electrodes (width of the working potential range), and the possibility of renewing the electrode surface by simply removing the outer layer. A shortcoming of graphite electrodes is the high residual current /165/, which is due to the reduction of oxygen in the pores /168/ and adsorbed oxygen /166/. However, this shortcoming can be eliminated by specially preparing the material used /70, 105, 167–172/.

The most widely used electrodes in stripping voltammetry are prepared from spectrally pure graphite impregnated with an epoxy resin and polyethylenepolyamine /81, 105, 171/. The author and coworkers have used these electrodes to study the deposition and stripping of several metals /172/ and insoluble compounds /102/, described methods of determining many elements, some of which are presented in the Appendices, and proposed using these electrodes for the analysis of industrial solutions /173/.

Perone used a wax-impregnated graphite electrode for the deposition and stripping of silver /174/ and mercury /75/. Later, Perone and Stapelfeldt /82/ proposed this electrode for use in differential stripping voltammetry. Roizenblat and Brainina /171/ used electrodes impregnated with paraffin and

a paraffin-polyethylene mixture in alternating-current anodic
stripping voltammetry of metals.

The preparation and properties of nonporous graphite, pyrolytic /175, 176/ and vitreous carbon /177—179/ have been reported. The anodic dissolution of copper from the surface of
a pyrolytic graphite electrode has been studied by Vassos and
Mark /180/.

Another type of carbon electrode is an electrode from a
carbon paste which is a mixture of powdered charcoal with an
organic nonvolatile liquid not miscible with water /181—183/.
The electrochemical reduction and dissolution of various metals
/184/ from the surface of this electrode, especially for gold
and silver /67, 68, 185/, as well as manganese /186/, have been
studied. However, the use of the paste electrode is limited due
to the oxidation of the organic component of the electrode /67/.
In the potential range for the reduction of oxygen on the paste
electrode, there is always a current rise even in a deaerated
solution, which hinders use of the electrode in the cathodic
potential range.

Monen, Specker, and Zinke /70/ carried out a comparative
investigation of various vitreous and pyrolytic carbon electrodes
from spectrally pure charcoal impregnated with wax and an
electrode from a carbon paste for the determination of 1 to
100 ng/ml Ag^+ in an 0.2 M potassium nitrate solution. Fairly
reproducible results were obtained by using the carbon paste
electrode* and the wax-impregnated electrode whose surface
was restored after each measurement. The amount of silver
deposited on the pyrolytic graphite electrode and the vitreous
carbon electrode decreased with repeated measurements. The
authors assume that this effect is due to adsorption phenomena
and the decrease in the number of active centers on the electrode
surface. Unfortunately, the electrodes mentioned in this reference cannot yet be considered sufficiently studied. The data
in the literature on their use are still not very abundant.

The graphite electrode impregnated with an epoxy resin and
polyethylenepolyamine has been most widely investigated.
According to the results reported in many works this electrode
provides fairly reproducible results. In most cases different
graphite electrodes are interchangeable. The difference

---

* In the potential range for the discharge—ionization of silver on the paste electrode,
  secondary electrode processes do not occur.

between individual electrodes appears under high anodic or cathodic polarization. In the first case, especially in acid medium, the residual current is less on electrodes impregnated with a paraffin-polyethylene mixture, and in the second, on electrodes impregnated with a resin and polyethylenepolyamine.

There are applications in stripping voltammetry for solid electrodes coated with mercury, known as film electrodes. The mercury coating makes this method similar to amalgam polarography with accumulation. The amalgamated electrodes, especially the silver amalgamated electrodes used in amalgam polarography with accumulation /189/, may be considered a variety of film electrodes. However, in the strict sense of the word, electrodes composed of a material not dissolving in mercury on which a transition amalgam layer of variable thickness does not form should be called mercury film electrodes. Mercury film electrodes composed of platinum coated with mercury /190, 191/ for the determination of Pb^{2+}, Cl^-, and I^- ions have been proposed /190, 191/. Roe and Toni /192/ derived equations describing the dissolution of a metal from the mercury film electrode and verified the derivation experimentally using a nickel electrode coated with mercury.

The mercury-graphite electrode with a polished surface for anodic voltammetry has been described by Matson, Roe, and Carriett /193/. They used spectrally pure graphite impregnated with paraffin in a vacuum. Hg^{2+} was introduced into the test solution to a concentration of $5 \cdot 10^{-8} - 1 \cdot 10^{-7}$ g-ion/liter, and the electrode was polarized for 10 min at -0.2 V. Microscopic analysis showed that approximately a tenth of the mercury was on the surface in the form of drops with a diameter of about 0.01 mm. The remainder was deposited in the form of far smaller drops.

Perone et al. /194, 195/ used graphite electrodes with a mercury coating in various types of voltammetry and chronopotentiometry. Roizenblat and Brainina /85/ proposed making such an electrode directly in the concentrating of a metal by adding a certain amount of mercuric oxide nitrate to the sample solution.

The mercury-graphite electrode is recommended in those cases in which the metals deposited in the concentrating step form solid solutions or intermetallic compounds /85/. In the presence of mercury, the interaction between the metals is

weakened, and the difficulties arising in the simultaneous determination of the concentration of the ions of several metals are
considerably reduced. The nature of the mercury-graphite
electrode has not been clarified. Apparently, these electrodes
are transitional between solid and amalgam electrodes.

TABLE. Mutual solubility of metals and mercury

Metal	Concentration in mercury-graphite electrode, wt.%	Solubility, wt.%		Temperature, °C	Reference
		metal in mercury	mercury in metal		
Copper	0.025—2	0.002		20	218
Lead	0.02—2	1.1	23	0	
		2.7	23	50	87
Cadmium	0.02—2	3	33	0	
		10	34	50	87
Antimony	0.05—0.02	$2.9 \cdot 10^{-5}$		20	87
Bismuth	0.03—1	1.4		18	87
Tin	0.5	0.56		20	87
					219
Thallium	0.5—1	14—18	18	0	
		42	16.5	25	87

The table compares the mutual solubility of metals and
mercury with the approximate concentration of the metal in the
mercury-graphite electrode which is obtained during analysis.
According to the table, the concentration of metal varies over
a wide range. The upper limit of this range sometimes coincides with or exceeds the solubility of the metal in mercury.
Even here, no interaction is observed between metals on the
electrode in the presence of mercury. Apparently, mercury
prevents the construction of a common crystal lattice for the
metals being deposited on the electrode even in amounts insufficient for the formation of a liquid amalgam.

These electrodes are indifferent to the electrode process
studied and are used in stripping voltammetry of metals and
variable-valence ions. In the cathodic stripping voltammetry
of anions, electroactive electrodes, silver /154/ and stationary
mercury /145, 147, 150, 153, 155—157/ electrodes are employed.
Methods for preparing the most frequently used electrodes
are given below.

Graphite electrode (type I). First, 10–15 g epoxy resin ED-6 is heated, an equal volume of acetone is added, the mixture is stirred, and 0.2 ml polyethylenepolyamine is introduced. The mixture is transferred to a thick-walled test tube bent at a right angle and heated to remove the acetone. Rods made from spectrally pure carbon with a diameter of 2 mm and a length of 10–15 mm are placed in the horizontal part of the test tube. Air is pumped out of the system, the rods are dropped into the resin and impregnated in vacuum with heating for 6 hr. The impregnated rods are left in the air until the resin hardens completely, are then subjected to heat treatment at 100–120°C for 6 hr, and the ends are sheared off. Each rod is inserted into a glass tube, a layer of polyethylene powder is applied to the lateral face of the electrode and the adjacent position of the tube end fused. This operation is repeated four or five times.

Electrical contact is made through mercury or carbon powder.

Graphite electrode (type II) /183/. First, 50 g carbon powder is mixed with 20 ml of a hydrophobic organic liquid. A small amount of the paste is press-fitted into a glass tube.

A special study has shown that the conditions for preparing the paste must be rigidly standardized since the stripping current is a function of the size of the carbon particles and the ratio of liquid and solid phases /196/. A shortcoming of this electrode is the high residual current at potentials more negative than -0.2 V. This is apparently due to the reduction of oxygen adsorbed on the finely ground carbon /187/.

Graphite electrode (type III). A low pressure polyethylene powder (25–30 wt. %) is introduced portionwise into melted paraffin, and the mixture is heated to a temperature not above 100°C until a homogeneous mass forms. The mixture is transferred to a right-angle test tube, and rods made from spectrally pure graphite ($l = 10$–15 mm, $d = 2$ mm) are placed in the horizontal part. The air is pumped out, then the electrodes are dropped into the mixture, the system is held in a vacuum for 6 to 8 hr after placing the test tube in a glycerin bath at 110–120°C. Then, the electrodes are withdrawn and cooled, and each electrode is inserted into a glass tube after its tip is cut off. The lateral surface of the electrodes and the adjacent end of the tube are coated with the paraffin-polyethylene mixture.

Electrical contact is made through mercury or powdered charcoal.

Pyrolytic graphite electrode. A mixture of natural gas and nitrogen (1:3) at atmospheric pressure is passed at a rate of 450 ml/min over a ceramic rod support in a rotating furnace at 1025°C. The support is previously carefully washed and dried. The system is cooled in a current of nitrogen /197/. To produce volume electrodes of pyrolytic or vitreous carbon, samples obtained by an industrial method are used.

Hanging mercury drop electrode. Mercury is deposited on a platinum wire with a diameter of 0.2 mm and a length of 0.1– 0.2 mm inserted into a glass tube. A large platinum electrode serves as the counter electrode.

The electrodes are placed in 1 N nitric acid, the working electrode undergoes anodic polarization by a current of 20 mA for 1 min. The polarity is reversed, and the polarization is repeated with a current of 20 mA for 1 min. The electrodes are then transferred to a mercuric nitrate solution, and mercury is deposited at a current of 30 mA for one minute. The electrodes are washed with water, and the mercury drop is cathodically polarized in 1 N nitric acid by a current of 20 mA for 1 min.

The last step should be repeated after each determination.

CHOICE OF EXPERIMENTAL CONDITIONS

ELECTROLYSIS PROCEDURE

Depending on the problem, analysis of a solution or an insoluble compound, the electroactive substance is either electrochemically concentrated on the electrode surface by prior electrolysis or mechanically introduced into the body of the electrode.

The analysis of a solution consists of three steps: 1) electrodeposition, 2) a settling period, and 3) electrochemical dissolution. The electrode potential is a fundamental parameter of the first step. It is selected in the region of the limiting diffusion current of the ions being concentrated (the optimum potentials are given in the Appendix). In the simultaneous determination of the ion concentration of several metals, the electrolysis potential is selected in the limiting diffusion current range corresponding to the most negative potential.

Shifting the electrolysis potential in the negative direction, if it corresponds to the limiting current, does not alter the nature of the polarization curves. An exception is the discharge of antimony (III) ions. In this case, there is a potential range in which the concentrating of antimony is slowed down /77/. This is probably due to repression of the cathodic process when the sign of the electrode surface charge is changed. A similar phenomenon should be observed in all similar cases, such as in the discharge of indium (III) ions /198/. In other cases, this effect is not detected. At the potential for the limiting diffusion current, the amount of deposit formed on the electrode is proportional to the concentration of the ions participating in the process and the electrolysis when the stirring of the solution is uniform or a rotating-disk electrode is used.

The settling step is carried out either at the electrolysis potential or at a potential close to the equilibrium value with the purpose of decreasing the current for the reduction or oxidation of the background or dissolution of impurities.

The stripping is performed in different ways. Most commonly, a linear variation of the potential with time is employed, and the dependence of the current on the electrode potential or on the time is recorded. Either the maximum current (anodic or cathodic depending on the nature of the electrode reaction) or the number of coulombs is measured. The value of the maximum current is determined uniquely by the concentration of the ions involved in the process. Therefore, a calibration graph may be used to determine an unknown concentration. Such a graph is necessary if the dependence of the maximum stripping current on the concentration of the sample ions is not directly proportional. If the calibration graph is a straight line arising from the origin, use of a known addition is possible. The choice of the electrode potential scanning rate is important. Increasing the scanning rate increases the sensitivity, since the maximum current is directly proportional to the electrode potential scanning rate. However, the double layer charging current which interferes with the measurements then increases and the resolution decreases. The optimal potential scanning rate usually is 0.01–0.2 V/sec. Sometimes, the dissolution step is carried out with a linearly varying potential with a superimposed variable component /171, 199/, and the amplitude of the alternating current is measured.

In principle all the methods for carrying out this step which
are used in stripping voltammetry with the mercury electrode
may be employed, namely, chronoamperometry, chronopotentiometry, chemical dissolution, and dissolution under constant
potential with subsequent reduction (or oxidation) under a linearly
varying potential.   These methods have been described in detail
by Barendecht /200/ and there is no sense in dwelling on them
any further since their use in stripping voltammetry is still rare.

In the analysis of insoluble substances, the electroactive compound is previously introduced into the electrode paste.   The
stripping process in this case is similar to that described.

REPRODUCIBILITY AND VALIDITY OF
THE RESULTS

The main sources of erros are the adsorption of surface-
active substances on the electrode which reduces the stripping
currents with time, interaction of metals on the electrode surface which distorts the polarization curves, unstable hydrodynamic conditions, and adsorption of hydrogen and oxygen on
the electrode surface.

To obtain reproducible results, traces of surface-active substances are removed from the solutions or the electrode surface is renewed after a certain number of measurements /202/.
In those cases in which the effect of surface-active substances
is especially noticeable, such as in the study of electrode process
kinetics, the solutions and water used are filtered into the electrolytic cell through an activated charcoal column.   The charcoal
is prepared by a method recommended by Cornish et al. /144/
or obtained by charring lump sugar in the air and calcining it
at red heat in a current of an inert gas.   In analytical applications, the effect of surface-active substances is not so important.   Special preparation of the solutions is usually required
after separations carried out with ion exchange resins.   In
these cases, the eluate is evaporated until fairly dry and treated
several times with nitric and hydrochloric acid.   If the sample
solution contains surface-active substances for whatever
reason, various methods of mechanically cleaning the electrode
such as scraping off the upper layer of graphite, or friction on
an abrasive are employed.   The use of an abrasive /173, 203/

allows use of anodic stripping voltammetry in industry for prolonged automatic analysis of solutions containing a large number of surface-active substances. For work with a hanging mercury drop electrode, periodic cathodic polarization of the electrode gives good results.

The interaction of deposits on an indifferent electrode appears in the formation of solid solutions and intermetallic compounds. In these cases, mercury-graphite electrodes should be used /85, 203/. The mercury may be applied to the electrode beforehand or at the same time as the metal is deposited. The latter method is preferable since mercury inhibits formation of a common crystal lattice for the metals. As a result, each metal in the anodic cycle is dissolved in its own potential range, the interaction of the metals does not appear, and thus the validity of the analysis results is assured.

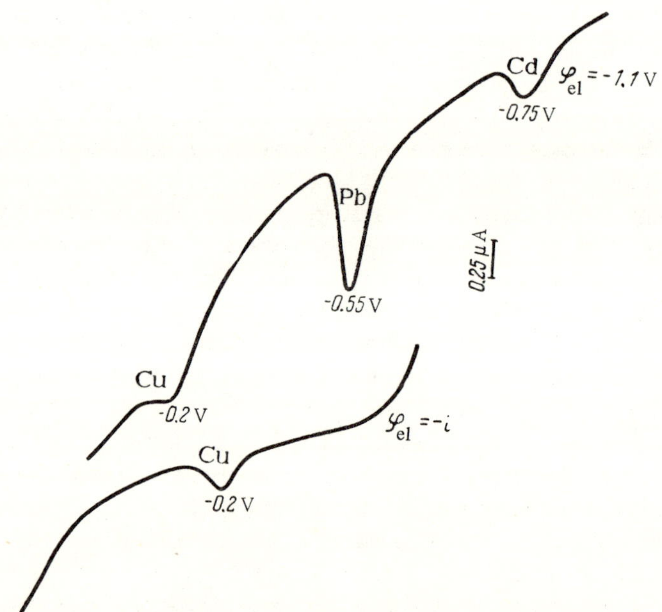

FIGURE 63. Effect of the adsorption of hydrogen on the polarization curves for the stripping of copper, lead, and cadmium obtained after the electrolysis of a solution 0.5 N in HCl, $5 \cdot 10^{-9}$ M in $CuCl_2$, $5 \cdot 10^{-9}$ M in $PbCl_2$, $5 \cdot 10^{-9}$ M in $CdCl_2$, and $5 \cdot 10^{-6}$ M in $Hg(NO_3)_2$ for $\tau_1 = 10$ min.

Sometimes, the electrolysis of acidic solutions at potentials more negative than -1.0 V is accompanied by the adsorption of hydrogen on the electrode surface.   Removal of hydrogen is achieved reversibly in the -0.4 to 0.1 V potential range.   The anodic current observed in this region, as evidenced from Figure 63, distorts the polarization curve and hinders measurement of the currents of electropositive metals such as copper.   In this case, all the metals being determined are first concentrated at an adequately negative potential, and the oxidation currents of cadmium and lead are measured by recording the anodic polarization curve.   Then the electrolysis is repeated at a more positive potential, and the oxidation current for electropositive metals such as copper is found.

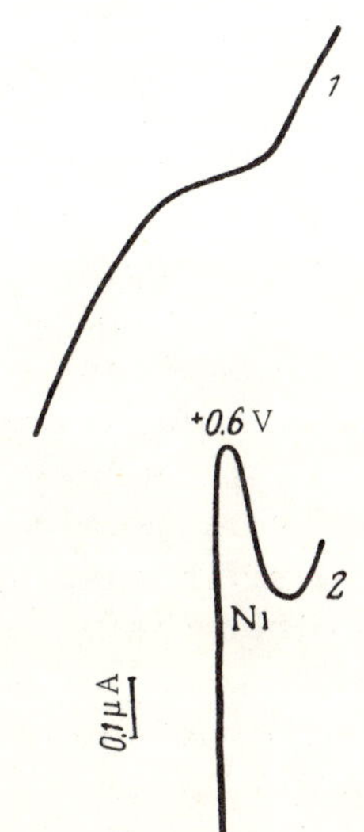

FIGURE 64. Effect of oxygen adsorption on the polarization curves for the reduction of the nickel-dimethylglyoxime compound obtained after the electrolysis of a solution 0.05 N in KOH, $1 \cdot 10^{-4}$ M in dimethylglyoxime, and $7 \cdot 10^{-8}$ M in $NiSO_4$ for $\tau_1 = 30$ sec and $\varphi_{el} = 0.8$ V.

Under a fairly high anodic polarization, especially in an alkaline medium, oxygen is apparently adsorbed on the electrode, and its reduction current often masks the current from the electrode process of interest (Figure 64).   In those cases in which the deposit concentrated on the electrode does not dissolve in the cathodic cycle, such as the compound of nickel and dimethylglyoxime, this difficulty is easily eliminated if the first polarogram (curve 1) is discarded, and the cathodic current is measured from the second (curve 2) or a subsequent curve.

The preparation described for electrodes and solutions enables us to obtain results for concentrations ranging from $5 \cdot 10^{-9}$ to $1 \cdot 10^{-5}$ g-ion/liter with an error not exceeding 10—20 rel. %.

DISTORTIONS DUE TO OHMIC VOLTAGE DROP

An ohmic voltage drop in the solution and the intrinsic potential of the electrode cause distortion in the linearity of the potential scanning with time, especially in the dissolution of macrophases. This distorts the polarization curve and results in a decrease in the maximum current. Polarization curves for the stripping of silver (Figure 65) and thallium hydroxide (Figure 66) have been obtained without compensating for the ohmic voltage drop (two-electrode system) and with compensation for the ohmic voltage drop (three-electrode system). Polarization curves 1, which were obtained without compensating for the ohmic voltage drop, tail off along the potential axis; the maximum current is lower in this case than for the stripping of the same amount of deposit with compensation for the ohmic voltage drop (curves 2). The distortions in the polarization curves are due to the fact that, without compensation, the potential is not a linear function of the time. The use of the three-electrode system with the polarograph model 02-TsLA usually allows us to obtain a linear potential scan if the current passing through the cell does not exceed 50 to $70 \, \mu A$. In other cases, a potentiostat should be used as the power supply.

In a kinetic study of electrode processes, direct proof must be obtained showing that the electrode potential is in fact a linear function of time in the stripping process. Therefore, the change in the potential of the working electrode relative to an independent reference electrode is recorded together with the polarogram using an oscillograph or a recording potentiometer with a high resistance input.

Dissolution of microphases $(i < 10 \, \mu A)$ is not accompanied by marked distortion in the linearity of the voltage scanning, and the polarization curves obtained from the two-electrode and three-electrode systems are identical. During an analysis small amounts of deposits, whose dissolution currents are low, usually form, and in these cases we may use a two-electrode system without compensating for the ohmic voltage drop.

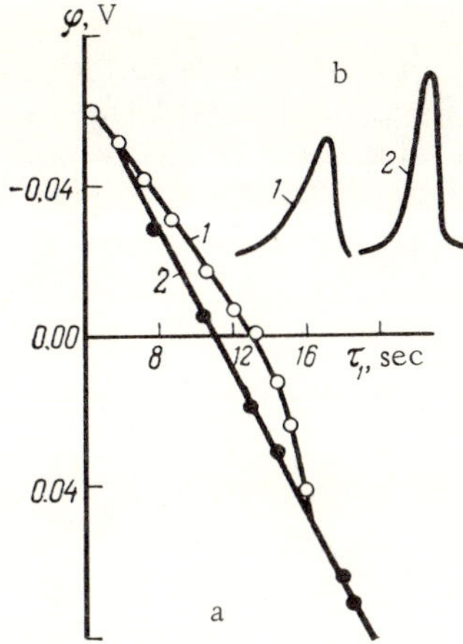

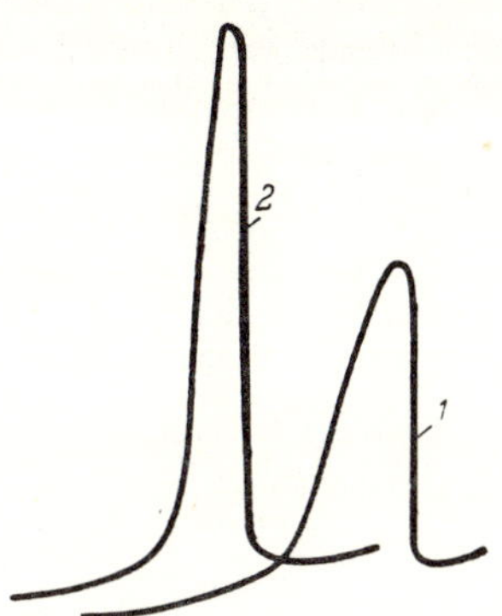

FIGURE 65. Change in time of the electrode potential (a) and polariza-tion curves (b) for the stripping of silver deposited for φ_{el} = -0.6 V and τ_1 = 1 min from a solution 1 N in KNO_3 and $3 \cdot 10^{-5}$ M in $AgNO_3$:

1) two-electrode system; 2) three-electrode system.

FIGURE 66. Polarization curves for the stripping (v = 0.04 V/sec) of thallium (III) hydroxide deposited for τ_1 = 1 min and φ_{el} = 1.2 V from an ammonia solution (pH = 7), 0.35 M in $(NH_4)_2SO_4$ and $2 \cdot 10^{-4}$ M in Tl_2SO_4:

1) two-electrode system; 2) three-electrode system.

Recording the polarization curves for the dissolution of larger amounts ($c \geqslant 1 \cdot 10^{-6} M$, τ_1 = 1 min) without compensating for the ohmic voltage drop causes the dependence of the maximum current on the amount of substances on the electrode or on the concentration of the ions being determined to be linear in contrast to the theoretical curve (see Chapter II). When a calibration graph is used, this behavior does not introduce an error in the results of the analysis.

Compensating for the ohmic voltage drop is necessary in an analysis when the elements being determined have close oxida-tion potentials and are present in the solution in noncommensur-ate amounts. On the polarization curve for the oxidation of lead and cadmium, two current maxima are observed which do not differ from the values corresponding to the individual metals at ratios of [Cd]:[Pb] $\leqslant$ 1:100 and [Cd]:[Pb] = 40:1. When the cadmium content in the deposit is increased further, the current for the stripping of lead masks the anodic current for cadmium when the polarization curve is recorded with a two-electrode polarograph system without compensating for the ohmic voltage drop. If the stripping of cadmium is carried out at a constant potential or with compensation for the ohmic voltage drop (with a three-electrode system), the current for the stripping of lead remains unchanged with an eighty-to a hundredfold cadmium excess. The polarization curve distortion in the first case is due to nonlinearity in the electrode potential change with time for large currents.

Chapter VI

PROSPECTS FOR ANODIC AND CATHODIC STRIPPING VOLTAMMETRY

APPLICATION OF THE METHOD IN AUTOMATIC ANALYSIS

To achieve automation, we may introduce the sample solution and nitrogen for expelling oxygen into an apparatus (Figure 67) consisting of vessel 3, measuring device 2 for the required reagents, and electrolytic cell 4 with a graphite electrode for the concentrating of metals and variable-valence ions or with a hanging mercury drop electrode for the concentrating of anions. An abrasive rod used for mechanically cleaning the graphite electrode is glued into the bottom of the electrolytic cell. Mechanical cleaning is carried out automatically in response to a signal from the control device after every one to ten measurement cycles. The continuous polarization of the electrode and the automatic recording of the polarization curves are performed /173, 203/ by an 02-TsLA oscillographic polarograph.

Automating the three methods described in Chapters II through IV is discussed below.

ANODIC STRIPPING VOLTAMMETRY OF METALS

In the determination of the ion concentration of individual metals and metals not interacting with each other, we may use a graphite electrode whose surface is periodically renewed mechanically. In the analysis of metals forming solid solutions or intermetallic compounds, mercury is deposited on the electrode either by prior electrolysis of an acidic solution of mercuric nitrate or directly in the concentrating step for the metals being determined (a divalent mercury salt is introduced into the test solution). This mercury-graphite electrode is distinguished from the mercury film electrode /189/, employed in amalgam polarography with accumulation, by the ease of its preparation and the fact that the small amount of mercury deposited on the graphite hardly alters the large working range of potentials for the graphite electrode. Only the 0.2—0.3 V interval, in which the electrochemical dissolution of mercury occurs, is excluded from this range.

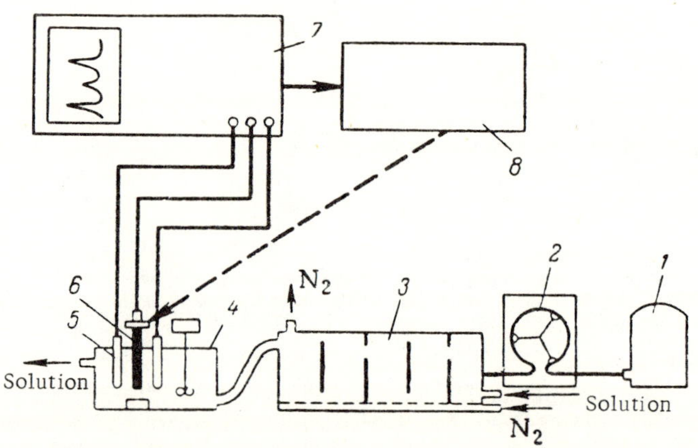

FIGURE 67. Apparatus for automatic determination of the ion concentration of several elements by stripping voltammetry:

1) tank for reagents; 2) measuring device; 3) vessel for mixing the test solution with the reagents; 4) electrolytic cell; 5) saturated calomel counter electrodes; 6) graphite electrode; 7) OP-02TsLA polarograph; 8) control device for mechanical cleaning of the electrode.

There are two possible modes for using the graphite electrode with a previously formed mercury film in continuous automatic measurements. The first mode involves limitation of the potential in the anodic cycle to a value insufficient for the dissolution of mercury. Good results are obtained in the determination of electronegative metals such as lead and cadmium. The oxidation current of copper increases with time, since copper does not dissolve at such potential limitation. Decreasing the voltage scanning rate to 0.02 V/sec in the measuring step eliminates this phenomenon. The second possibility arises for

the analysis of solutions in which the mercurous salts are insoluble. Here, the anodic polarization in the measuring step is
accompanied by the formation on the electrode surface of an insoluble mercury salt which is reduced under cathodic polarization in the concentrating step and a usable mercury layer is
regenerated. Figure 68 presents the anodic polarization curves
for the oxidation of cadmium, lead, copper, and mercury and the
cathodic curve for the reduction of mercurous chloride, which
forms in the anodic cycle. The presence of a hysteresis enables
us the repeated use of an electrode with a once formed mercury
film. The coefficient of variation does not exceed 10%.

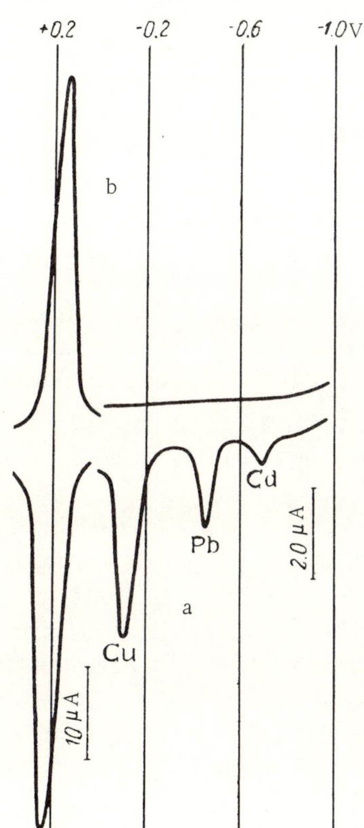

FIGURE 68. Polarization curves ($v = 0.02$ V/sec)
for the oxidation (a) of cadmium, lead, copper, and
mercury deposited for $\varphi_{el} = 1.0$ V and $\tau_1 = 2.5$ min
and the reduction curve (b) for mercurous chloride
formed in the anodic cycle in a 1 N KCl supporting
solution.

To obtain an electrode with a renewable mercury film,
$Hg(NO_3)_2$ is introduced into the solution (its concentration should
be $1 \cdot 10^{-5}$ to $1 \cdot 10^{-4}$ M). The mercury film forms on the

graphite electrode surface directly in the concentrating step and is removed in the measuring step. This variant is obviously most preferable, especially in the analysis of industrial solutions containing surface-active substances, since it permits mechanical cleaning of the electrode. The coefficient of variation in the determination of 10^{-7}–10^{-6} g-ion/liter Cu^{2+}, Pb^{2+}, and Cd^{2+} in industrial solutions of $ZnSO_4$ is 7 to 10%.

STRIPPING VOLTAMMETRY OF VARIABLE-VALENCE IONS

Automation of this type of stripping voltammetry has been demonstrated for the continuous concentrating and determination of thallium (I) in the form of $Tl(OH)_3$ and nickel (III) as a dimethylglyoxime compound. A graphite electrode is used as the indicator electrode.

Thallium is concentrated on the electrode as a result of the reaction $Tl^+ + 3OH^- \rightarrow Tl(OH)_3 + 2e$, which occurs at a potential of $+1.0\,V$. The cathodic polarization of the electrode is accompanied by the complete dissolution of the deposit, after which the electrode may be used for subsequent measurements. An ammonium sulfate solution containing Tl^+ ions is analyzed. Introduction of additional reagents and mechanical cleaning of the electrode are not required. All steps are carried out automatically.

Nickel is concentrated on the electrode as a result of a reaction which occurs at $+1.0\,V$ and may formally be described by the equation $Ni(HD)_3^- + 2OH^- \rightarrow l + Ni(OH)_2DH + 2HD^-$. Automatic analysis of this type is more difficult since introduction of special reagents into the sample solution is required. The solution must be 0.02 M in KOH and $1\cdot10^{-5}$ M in H_2D. In addition, reduction of the deposit formed on the electrode in the concentrating step is not accompanied by its dissolution. In each subsequent concentrating cycle an additional amount of precipitate is deposited, and the reduction current rises accordingly, without reflecting a change in the concentration of ions in the solution. Therefore, for the determination of nickel, a measuring device is used to introduce the necessary reagents into the flowing test solution, and the electrode surface is cleaned mechanically after each measurement.   All the operations are also performed automatically.

The coefficient of variation in the determination of 10^{-6} to 10^{-5} g-ion/liter Tl^+ in a solution 0.5 M in $(NH_4)_2SO_4$ and $1 \cdot 10^{-7} - 2 \cdot 10^{-7}$ g-ion/liter in Ni^{2+} in a solution 1 N in KNO_3, 0.05 N in KOH, and $1 \cdot 10^{-5}$ M in H_2D does not exceed 8—10%.   The accumulation time is 3 min.

CATHODIC STRIPPING VOLTAMMETRY
OF ANIONS

Usually a hanging mercury drop electrode is used as the indicator electrode here.   The main difficulty in carrying out the automatic determination of anions with the hanging mercury drop electrode is associated with the adsorption of traces of surface-active substances usually present in solutions.   This effect shows up most vividly in the determination of small amounts of Cl^- $(5 \cdot 10^{-6} - 1 \cdot 10^{-5}$ g-ion/liter) and, to a lesser extent, in the determination of Br^-, I^-, and S^{2-}.   The Cl^- ions are concentrated /203/ from a 1 N KNO_3 nitric acid solution (pH= 1) containing $6 \cdot 10^{-6}$ to $2 \cdot 10^{-5}$ g-ion/liter Cl^- onto the hanging mercury drop electrode at a potential of -0.05 V relative to a mercurous sulfate electrode for 70 sec.   To eliminate the effect of surface-active substances the electrode is polarized at from -0.05 to -1.6 V in the stripping step and the hydrogen evolved destroys the adsorption film.   The coefficient of variation does not exceed 8%.   In the analysis of substances in whose presence adequate cathodic polarization is impossible, the electrode should be transferred to an auxiliary acidic solution by using the electrolytic cell developed by Tur'yan /204/.

Therefore, an automatic determination consists of two or three steps depending on the nature of the deposit which forms on the electrode: 1) polarization of the electrode at a constant potential (concentrating), 2) linear scanning of the voltage with time (measuring), and 3) mechanical renewal of the electrode surface.   Another procedure for the varying of the electrode potential with time may be used, such as polarization by voltage pulses and measurement of the current pulse.   The third step is required in systems for which hysteresis is found or in the presence of a large amount of surface-active substances.   In other cases, the preparation of the electrode for the following cycle is accomplished directly during the measurement.

APPLICATION OF ALTERNATING CURRENT IN STRIPPING VOLTAMMETRY

Alternating current is still hardly used in stripping voltammetry, especially in those variants which employ solid electrodes. The main shortcoming of solid electrodes, which shows up in alternating current measurements, is the high capacitance of the electric double layer caused by the roughness of the surface and the resulting high residual current. Thus, graphite electrodes impregnated with an epoxy resin (type I), which are widely and successfully employed in direct-current stripping voltammetry, are hardly suitable for work with a vector polarograph /171/. The capacitance current observed on electrodes impregnated with paraffin in vacuum are significantly lower. However, the results of the measurements are not sufficiently reproducible.

In vector polarography, graphite electrodes impregnated in vacuum with a paraffin-polyethylene mixture (type III) can be employed successfully /171/.

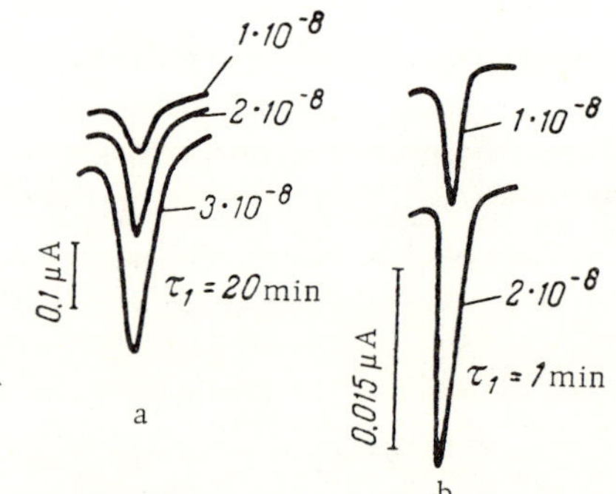

FIGURE 69. Direct-current polarograms (a) and vector polarograms (b) for the stripping of lead deposited at $\varphi_{el} = -0.9\,V$ from 1 N HCl solutions with different Pb^{2+} concentrations (numbers on the curves: g-ion/liter).

Polarization curves for the stripping of lead are presented in Figure 69. According to this figure, the vector polarograms are well defined and a signal suitable for measurement is obtained for a concentrating time twenty times less than for the recording of a direct current polarogram. The maximum of the active current component for the stripping of copper, lead, and cadmium in the range studied is directly proportional to the concentration of the respective ions in the solution, the deposition time, and the rate of varying the direct component of the potential.

Under similar conditions in alternating-current amalgam polarography with accumulation, the dependence of the maximum current on the electrolysis time /111, 205/ is nonlinear, and in several cases the current maximum decreases with increasing deposition time. This reverse dependence is explained by the decrease in the rate of the electrode process as a result of the adsorption of surface-active substances during the electrolysis. The linear nature of this dependence in stripping vector polarography with accumulation on a graphite electrode implies that the sensitivity of this method may be increased significantly as a result of prolonged prior electrolysis.

Kaplan and Rezakova /199/ developed a method for the determination of selenosulfate and tellurite involving the concentrating of selenium as mercurous selenide and tellurium as the element on a hanging mercury drop electrode with the subsequent recording of the cathodic alternating-current polarograms.

Concentrating selenium as copper selenide on the hanging mercury drop electrode has also been demonstrated /206/. Copper selenide forms in the electrolysis of a selenious acid solution in the presence of a copper excess. There is a significant increase in the sensitivity of the determination of selenium in comparison to that obtained in the concentrating of selenium as mercurous selenide.

The authors showed that the maximum in the variable component of the stripping current for the deposit is proportional to the selenide ion concentration.

Pats and Semochkina /207/ proposed a vector polarographic method for the determination of selenium in tellurium and ores involving the prior concentrating of selenium as mercurous selenide on a hanging mercury drop electrode.

Pats, Vasil'eva, and Semochkina /208/ described the determination of 10^{-6} to $10^{-4}\%$ tellurium in selenium on a background of hydriodic acid.   The tellurium is accumulated on a hanging mercury drop electrode, and then the cathodic vector polarogram recorded.   The sensitivity of the determination is $2 \cdot 10^{-9}$ M.

STRIPPING VOLTAMMETRY IN THE BODY OF THE ELECTRODE

One of the first attempts to introduce a solid electroactive substance into the volume of a solid indifferent electrode was apparently made by Ruby and Tremmel /209/.   They prepared electrodes from a carbon paste containing silver chloride, bromide, or iodide, and investigated these electrodes by cyclic voltammetry, chronopotentiometry, and chronoamperometry. When the electrode was polarized cathodically, voltammetric curves with a current maximum were observed.   The investigations performed at constant potential showed that the flux of the electroactive substance into the reaction zone is an exponential function of the electrode potential.   The authors assumed that silver ions in the solid phase participated in the electrode process.

A similar electrode for the analysis of insoluble compounds was employed by Barikov, Rozhdestvenskaya, and Songina /201, 210/.   They mixed the finely ground sample to be analyzed with powdered charcoal prepared from spectrally pure charcoal roasted at 800°C.   As a binding material, α-bromonaphthalene (0.5 ml per gram of carbon) was used.   The authors called the electrode prepared from this mass a mineral-carbon paste electrode.   It has the properties of the substance contained in the paste and can be operated in the potential range of the carbon paste electrode.   The oxidation or reduction currents of the electroactive substance introduced into the paste depend on the degree of dispersion of the substance and the carbon, the concentration of the substance and the binding liquid in the paste, and the nature of the electrolyte.   These authors studied the oxidation of zinc, lead, and copper sulfides with three types of electrolytes for which 1) the anions of the electrolyte form a well-dissociated soluble compound with the cation of the sulfide, 2) the sulfide cation forms complex ions with the electrolyte, and 3) anions of the electrolyte form an insoluble compound with the cation of the sulfide.   In the first case, several curves which tail off along the potential axis were observed, in the second, the curve acquires a more symmetric shape with a well-defined maximum at a more negative electrode potential, and in the third, a sharp current peak, although of smaller magnitude, is observed.   The sulfide oxidation potential probably depends on the strength of the metal-sulfide bond in the crystal lattice and on the interaction energy of the cation liberated along with the electrolyte anion.   An electrolyte is selected in which zinc sulfide and lead sulfide are oxidized at different potentials.

The recording of currents for the electrochemical conversion of solids mechanically introduced into an electrode paste obviously opens up possibilities of determining different valence states of elements without dissolving the sample.

There is undoubtedly interest in the development of the electrode theory for complex pastes and its application to the analysis of different types of solids, in particular in the determination of ions of the same element with different valences.

INDIRECT METHODS OF ANALYSIS

The possibility of concentrating different organic compounds or their decomposition products on the surface of an electrode opens prospects of developing a method for the indirect determinations of metal ions.   The interaction of these ions with a reagent in the solution bulk decreases the equilibrium concentration of the reagent, decreases the amount of insoluble compound formed on the electrode surface in the concentrating step, and accordingly decreases its reduction current.

Cathodic stripping voltammetry of anions has been applied in the indirect determination of mercury, platinum, gold, and silver /211/.   Thiourea, thionalide, 2-mercaptobenzthiazole, and dithiooxamide were used as the reagents.   An indirect method for the determination of CN^- ions in a solution containing Cu^{2+} ions has been proposed.   The ions are concentrated as the insoluble copper (I) cyanide on the surface of a mercury electrode /212/.

The study of the oxidation of sodium diethyldithiocarbamate on the graphite electrode has shown that the electrolysis product is concentrated on the electrode and the deposit formed is capable of further oxidation and reduction. Both processes are described by characteristic curves with a current maximum (Figure 70) dependent on the electrolysis potential and directly proportional to the deposition time and the concentration of sodium diethyldithiocarbamate.

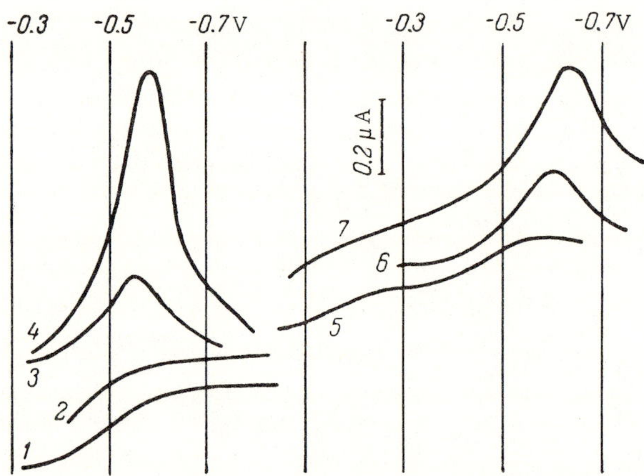

FIGURE 70. Polarization curves for the reduction (v = 0.011 V/sec) of the oxidation product of sodium diethyldithiocarbamate obtained after the electrolysis of a solution 0.1 M in NH_4OH, 0.1 M in NH_4Cl, and $6 \cdot 10^{-6}$ M in $C_5H_{10}NSNa$ under the conditions:

1) τ_1 = 2 min, φ_{el}= -0.3 V;   2) τ_1= 2 min, φ_{el}= -0.2 V;
3) τ_1 = 2 min, φ_{el}= 0;   4) τ_1 = 2 min, φ_{el}= 0.2 V;
5) τ_1 = 1 min, φ_{el}= 0 V;   6) τ_1 = 2 min, φ_{el}= 0 V;
7) τ_1 = 4 min, φ_{el}= 0 V.

This behavior allows development of an indirect method for the determination of the concentration of metal ions which interact with sodium diethyldithiocarbamate in the solution bulk. The authors observed a decrease in the maximum of the derivative of the current with respect to time in the electroreduction of a deposit concentrated from a solution into which a mercury salt was introduced. The decrease in $(di/dt)_{max}$ is proportional to the mercury concentration /131/.

This method is similar to that proposed by Berge and Jeroschewski /211/. A difference is that the latter determined the concentration of the reagent in the solution from the reduction current for the insoluble compound of the reagent with mercury ions which forms under anodic polarization of a mercury electrode in the sample solution.

Both methods may obviously be easily modified and employed as an amperometric titration with a considerably higher sensitivity than the conventional variant since the reagent is concentrated beforehand on the electrode.

CONCLUSION

Study of the electrochemical conversions of solids yields valuable information on the properties of electrode—deposit—solution systems. The behavior of small quantities of metals and chemical compounds on the electrode surface, the initial moments of crystallization, and the removal of the last layers from the surface of an indifferent electrode are of interest although they have been little studied so far. Several features of the processes occurring in microdeposit—electrode systems show up in the polarization curves for stripping of the deposit. The change in the slope of the curve describing the dependence of the maximum stripping current on the amount of deposit on the electrode with a linearly varying potential reflects the transition from a microphase to a macrophase and shows what amount of the substance on a given electrode is necessary for the onset of behavior of the corresponding macrophase. A comparison of electrochemical and radiographic studies enables us to understand and quantitatively describe the properties of such systems.

The nature of the polarization curves for the electrochemical dissolution of deposits is determined by the mechanism and kinetics of the electrode process.

At the present time, the kinetics of the stripping of silver and nickel from the surface of a graphite electrode and of mercury halide salts from the surface of a mercury electrode has been studied. In the future, the study of the kinetics of the stripping of other metals and compounds from the surfaces of other electrodes will probably not present any difficulties.

The proportional relationship between the current maximum for the electrochemical conversion of a substance localized on an electrode and the amount of this substance enables us to use the formation and dissolution of deposits as an integral microcoulometer for the current of a cumulative or single process. Examples of the former are systems in which there is hysteresis. A deposit once formed does not dissolve as a result of the reverse electrode reaction, such as the nickel-dimethylglyoxime system.

The dissolution of metals from the surface of a foreign-type electrode may be used for the determination of the thickness of electroplatings, and the characteristic features of polarograms for the oxidation of solid solutions and intermetallic compounds enable us to record and study electroalloys. By analyzing the polarization curves for the dissolution of deposits, useful information may be obtained on the electrophysical properties of the film and on several reactions in the solution. Thus, the nature of the dependence of the amount of compound formed on the electrode on the product $c_1^0 \tau_1$ enables us to judge the solubility of the compound and its passivating properties. Comparing the dependence of the amount of substance formed on the electrode on the concentration of electroactive ions and the electrolysis time, we easily find the concentration at which the ions in the initial valence state form an insoluble compound in the solution bulk, and thus evaluate the solubility of this compound. Recalling that the accumulation of cobalt (III) as a compound with 2-nitroso-1-naphthol on an electrode from $2 \cdot 10^{-5}$ M solution of the reagent ceases when the concentration of Co^{2+} ions in the solution is greater than $8 \cdot 10^{-7}$ g-ion/liter, we may conclude that the solubility product of the compound CoR_2 has a value of the order of 10^{-16}.

Hysteresis phenomena enable us to judge the mechanism of electrochemical conversions in solid phases.

The concentrating of variable-valence elements in the form of insoluble compounds on an indifferent electrode may apparently be used as a method for separating a number of ions from solution. The deposition of several elements in the form of hydroxides is used in analytical chemistry. There is also interest in the possibility of electrochemically synthesizing compounds containing organic groups.

Stripping voltammetry may be used /213–214/ to study the composition of complex ions and their redox reaction kinetics. There is a possibility of evaluating the rates of the electron transfer step, the chemical reaction, and the removal of ions from the electrode surface and to determine the composition of the complex ions involved in the electrode process.

The analytical possibilities of anodic and cathodic stripping voltammetry are obvious from the results cited in Chapters II through IV and summarized in the tables given in the Appendix. Comparative data on stripping voltammetry and other polarographic methods of determining several elements are presented

Comparison of different types of polarographic determination for several elements*

Ion determined	Classical polarography	Amalgam polarography with accumulation		Anodic stripping voltammetry of metals		Cathodic stripping voltammetry of anions		Stripping voltammetry of variable-valence ions			
	sensitivity, g-ion/liter	sensitivity g-ion/liter	interfering elements	sensitivity, g-ion/liter	interfering elements	sensitivity, g-ion/liter	interfering elements	sensitivity, g-ion/liter	interfering elements with ratio of interfering element to element being determined		
									$\leq$10:1	> 100:1	> 1,000
Ni^{2+}	$1 \cdot 10^{-5}$	—	—	$1 \cdot 10^{-7}$	Co, Cu, Pb, Sn in commensurate amounts			$1 \cdot 10^{-9}$	None	None	Co, Cu
Fe^{2+}	$1 \cdot 10^{-5}$	—	—	$1 \cdot 10^{-6}$	Pb, Cu in commensurate amounts			$1 \cdot 10^{-7}$	Co	—	—
Sb^{3+}	$1 \cdot 10^{-}$	$1 \cdot 10^{-10}$	Cu, Bi in commensurate amounts	$1 \cdot 10^{-8}$	Bi, Cu, Ag in commensurate amounts			$1 \cdot 10^{-8}$	I^{-} **	Bi	Sn, Pb
I^{-}	$(1-5) \cdot 10^{-5}$					$5 \cdot 10^{-8}$	Br^{-} in fiftyfold excess	$5 \cdot 10^{-9}$	Sb^{3+} **	Bi	Sn, Pb

* No data.

** There is a method of eliminating the effect of the interfering elements.

below.   The determination of nickel and iodine by concentrating these elements in the form of chemical compounds is considerably more sensitive and selective than polarographic methods employing other variants of the concentrating step.   The determination of antimony by stripping voltammetry of variable-valence ions is less sensitive than amalgam polarography with accumulation.   However, the stripping voltammetry method is recommended in the analysis of complicated mixtures containing elements such as copper, bismuth, tin, and silver.

The possibility of forming various compounds on the electrode as a result of cathodic and anodic reactions allows analysis of multicomponent systems without preliminary separation of the element being determined.   Depending on their chemical nature, these elements are concentrated as hydroxides, inorganic salts, ionic associates, or coordination compounds with organic reagents.

A very important advantage of an electrochemical concentrating method employing organic reagents over the corresponding chemical methods is the significant increase in the selectivity of the determination in the presence of elements which normally interfere with the analysis by a chemical method, since the selectivity of a reaction is increased by the individual electrochemical properties of the elements.

The development of stripping voltammetry will broaden the scope of the electrochemical determination of insoluble substances.

Thus, stripping voltammetry of solids is a valuable addition to currently employed physicochemical methods for research and analysis.

APPENDICES

I. STRIPPING OF A METAL FROM THE SURFACE OF AN INDIFFERENT ELECTRODE
/59, 60, 61/

General case

To solve equations (II.2)–(II.5) in Chapter II, we substitute for the independent variables:

$$\tau = \frac{nF}{RT}\,vt\;, \qquad \varkappa = \sqrt{\frac{nFv}{RTD}}\,x \tag{1}$$

and introduce the dimensionless functions

$$u(\varkappa,\tau) = \frac{c(\varkappa,\tau)}{a_\infty}\,, \qquad a' = \frac{a}{a_\infty} \tag{2}$$

Then, equations (II.2) – (II.5) become

$$\frac{\partial u}{\partial \tau} = \frac{\partial^2 u}{\partial \varkappa^2} \tag{3}$$

$$j = \frac{i}{nFSk_S\,a_\infty} = -\frac{D}{k_S}\sqrt{\frac{nFv}{RTD}} \cdot \frac{\partial u}{\partial \varkappa}\bigg|_{\varkappa=0} =$$

$$= a'\exp\left[\frac{\beta nF}{RT}(\varphi_1 - \varphi^0)\right]\cdot\exp(\beta\tau) - u(0,\tau)\exp\left[-\frac{\alpha nF}{RT}(\varphi_1 - \varphi^0)\right]\cdot\exp(-\alpha\tau) \tag{4}$$

$$a' = 1 - \exp\left[-\gamma\left(Q - Sk_S\,a_\infty\frac{RT}{v}\int_0^\tau jdt\right)\right]$$

$$u(l,\tau) = u^0$$

where l^+ is a distance from the electrode at which the concentration of Me^+ ions is assumed to equal the concentration in the solution bulk.

Let us represent equation (4) in the following manner:

$$f_1(\tau)\,u(0,\tau) - \left(\frac{\partial u}{\partial \varkappa}\right)_{\varkappa=0} =$$

$$= f_2(\tau)\left\{1 - \exp\left[-\gamma Q - A\int_0^\tau\left(\frac{\partial u}{\partial \varkappa}\right)_{\varkappa=0}d\tau\right]\right\} \tag{5}$$

Here

$$f_1(\tau) = \frac{k_S}{\sqrt{D \dfrac{nF}{RT} v}} \exp\left[-\frac{\alpha nF}{RT} (\varphi_1 - \varphi^0) - \alpha\tau \right]$$

$$f_2(\tau) = \frac{k_S}{\sqrt{D \dfrac{nF}{RT} v}} \exp\left[\frac{\beta nF}{RT} (\varphi_1 - \varphi^0) - \beta\tau \right]$$

$$A = \gamma S a_\infty \sqrt{\frac{RTDnF}{D}}$$

Equation (3) is approximated by the implicit difference equation

$$\frac{u_n^{m+1} - u_n^m}{\tau_1} = \frac{u_{n-1}^{m+1} - 2u_n^{m+1} + u_{n+1}^{m+1}}{h^2} \tag{6}$$

As the calculation grid, we select the set of points with co-ordinates $x_n = h/2 + nh$ $(h = l/N$, where N is an integer) and $\tau_m = m\tau_1$. For each moment in time τ_m, we introduce into the grid two fictitious points $x = -h/2$ and $x = l + h/2$ lying outside the section $(0, l)$. To determine the values of $u(x, \tau)$ at these points, boundary conditions are used. For a fixed n, relation (6) should be fulfilled for all interior points in the interval $(0 < x < l)$, i. e., for $n = 0, 1, \ldots, N - 1$. Thus, we easily obtain N equations which contain $N + 2$ unknowns: $u_{-1}^{m+1}, u_0^{m+1}, \ldots, u_N^{m+1}$. The two equations lacking may be obtained from the boundary conditions, which are approximated by the following relationships:

$$f_1^{m+1} \cdot \frac{u_1^{m+1} + u_0^{m+1}}{2} - \frac{u_0^{m+1} - u_{-1}^{m+1}}{h} =$$

$$= f_2^{m+1} \cdot \left[1 - \exp\left(-\gamma Q - A \sum_{i=0}^{m+1} \frac{u_0^i - u_{-1}^i}{h} \tau \right) \right]$$

$$\frac{u_{N-1}^{m+1} + u_N^{m+1}}{2} = u^0 \tag{7}$$

Equalities (7) together with equation (6) yield a system of $N + 2$ equations with $N + 2$ unknowns for a fixed m.

Let us write relations (6) and (7) in the following form (omitting the index $m + 1$ from the quantities u_n^{m+1}):

$$u_{n+1} + 2bu_n + u_{n-1} + d_n = 0 \tag{8}$$

$$F(u_{-1}, u_0) = 0 \tag{9}$$

$$u_N = x_{N-1} \cdot u_{N-1} + y_{N-1} \tag{10}$$

Here

$$b = 1 + \frac{h^2}{2\tau}; \qquad d_n = \frac{h^2}{\tau} u_n^m \tag{11}$$

$$x_{N-1} = -1; \qquad y_{N-1} = 2U_0 \tag{12}$$

The method of solving equations (8)–(10) consists in consecutively applying relationship (10) from right to left, in which the solution at the n-th point is obtained if it is known at the $(n + 1)$-th point. After obtaining a relation of the form

$$u_{n+1} = x_n u_n + y_n \tag{13}$$

and substituting it into equation (8), we obtain

$$u_n = \frac{1}{2b - x_n} u_{n-1} + \frac{d_n + y_n}{2b - x_n}$$

Assuming in this relationship that

$$x_{n-1} = \frac{1}{2b - x_n}; \qquad y_{n-1} = (d_n + y_n)x_{n-1} \tag{14}$$

we rewrite it in the following form:

$$u_n = x_{n-1} u_{n-1} + y_{n-1}$$

We solve the equations in the following manner:
1) we calculate d_n and b from equations (11);
2) we determine x_{N-1} and y_{N-1} from equations (12);
3) from equations (14) we find successively

$$x_{N-2}, \, x_{N-3}, \, \cdots, \, x_{-1}$$

$$y_{N-2}, \, y_{N-3}, \, \cdots, \, y_{-1}$$

4) solving the system

$$\begin{cases} F(u_0, u_{-1}) = 0 \\ u_0 = \varkappa_{-1} u_{-1} + y_{-1} \end{cases}$$

we calculate u_{-1};

5) we find successively

$$u_0, u_1, \ldots, u_N \text{ from equation (13).}$$

This problem was solved on a BECM-2M electronic computer from the implicit difference scheme by the consecutive application method /216, 217/.

II. IRREVERSIBLE ELECTROCHEMICAL DISSOLUTION OF A METAL

Derivation of formulas (11.10) —(11.14) in Chapter II
Into equation (II.9) we substitute the variables:

$$y = \exp\left[-\gamma\left(Q - \int\limits_0^t id\tau \right) \right] \tag{15}$$

$$y' = \gamma i \exp\left[-\gamma\left(Q - \int\limits_0^t id\tau \right) \right] \tag{16}$$

Then equation (II.9) becomes

$$\frac{y'}{y(1-y)} = k \exp\,(Bvt) \tag{17}$$

where

$$k = nFSk_S\, a_\infty \gamma \exp\left[\frac{\beta nF}{RT}\,(\varphi_1 - \varphi^0) \right]; \qquad B = \frac{\beta nF}{RT}$$

After integrating expression (17), we obtain

$$\ln\frac{Cy}{1-y} = \frac{k}{Bv}\,\exp\,(Bvt) \tag{18}$$

We find C from the condition $t = 0$, $y = \exp(-\gamma Q)$:

$$y = \cfrac{1}{C \exp\left[-\cfrac{k}{Bv} \exp(Bvt)\right] + 1} \tag{19}$$

where $\quad C = [\exp(\gamma Q) - 1] \exp\dfrac{k}{Bv}$

Differentiating equation (19) with respect to t, we obtain:

$$i = \frac{y'}{\gamma y} = \cfrac{kC \exp\left(Bvt - \cfrac{k}{Bv} \exp Bvt\right)}{\gamma \left[C \exp\left(-\cfrac{k}{Bv} \exp Bvt\right) + 1\right]} \tag{20}$$

After the values of C and B are substituted, equation (20) becomes (II.10).

For $\gamma Q \ll 1$ the exponentials may be expanded in series and limited to the first two terms. Furthermore, the first term in the denominator is much smaller than unity and may be neglected. Equation (20) then becomes the simpler relationship (21), which may also be obtained directly from equation (II.9) when $\gamma Q \ll 1$

$$i = kQ \exp\left[\frac{k}{Bv}(1 - \exp Bvt) + Bvt\right] \tag{21}$$

After the values of k and B are substituted, this expression becomes (II.11).

Differentiating expression (21) with respect to t, we find the condition for the maximum of the function

$$\frac{di}{dt} = kQ \exp\left[\frac{k}{Bv}(1 - \exp Bvt) + Bvt\right] \cdot (-k \exp Bvt + Bv) \tag{22}$$

$$\frac{RTSk_s a_\infty \gamma}{\beta v} \exp\left[\frac{\beta nF}{RT}(\varphi_1 - \varphi^0 + vt_{\max})\right] = 1 \tag{23}$$

Taking the logarithm of this expression and recalling the equality $\varphi = \varphi_1 + vt$, we easily obtain equation (II.12). Expression (II.13) is the result of substituting equation (23) into equation (21).

We obtain equation (II.14) by differentiating equation (22) with respect to t and setting the derivative equal to zero:

$$\frac{d^2 i}{dt^2} = kQ \left\{ \exp\left[\frac{k}{Bv}(1 - \exp Bvt) + Bvt \right] (-kBv \exp + Bvt) + \right.$$

$$+ (-k \exp Bvt + Bv)^2 \exp\left[\frac{k}{Bv}(1 - \exp Bvt) + \right.$$

$$\left. + Bvt \right] \right\} \left(\frac{k}{Bv} \exp Bvt \right)^2 - 3\frac{k}{Bv} \exp Bvt + 1 = 0 \tag{24}$$

Assuming $\dfrac{k}{Bv} \exp(Bvt) = \zeta$, we obtain the quadratic equation $\zeta^2 - 3\zeta + 1 = 0$, whose roots are $\zeta' = 0.4$ and $\zeta'' = 2.61$.   Hence

$$Bvt_{max} = \ln \zeta - \ln \frac{k}{Bv} \tag{25}$$

After the values of k and B are substituted, we obtain equation (II.14).

III. REVERSIBLE ELECTROCHEMICAL DISSOLUTION OF A METAL

Derivation of formulas (II. 21) — (II. 24)
Let us make a substitution of variables and combine the constants in equation (II.20)

$$y = \exp\left[-\gamma\left(Q - \int_0^t i\,d\tau \right) \right]; \qquad y' = y\gamma i$$

$$A = \frac{\delta}{nFSDc^0\gamma}[1 - \exp(-\gamma Q)]; \qquad \frac{nF}{RT} = B$$

Then expression (II.20) becomes

$$\frac{dy}{y(1 - y)} = \frac{1}{A} \exp Bvt \cdot dt \tag{26}$$

The solution of equation (26) satisfying the initial condition $y = \exp(-\gamma Q)$ at $t = 0$ may be represented as

$$y = \frac{1}{C \exp\left(-\dfrac{1}{ABv}\exp Bvt\right) + 1} \tag{27}$$

where

$$C = (\exp \gamma Q - 1)\exp \frac{1}{ABv}$$

We differentiate equation (27) with respect to t

$$\frac{dy}{dt} = \frac{C \cdot \dfrac{1}{A}\exp\left(-\dfrac{1}{ABv}\exp Bvt\right)\cdot \exp Bvt}{\left[C \exp\left(-\dfrac{1}{ABv}\exp Bvt\right) + 1\right]^2} \tag{28}$$

Hence

$$i = \frac{y'}{\gamma y} = \frac{C \cdot \exp Bvt \cdot \exp\left(-\dfrac{1}{ABv}\exp Bvt\right)}{A\gamma\left[C \exp\left(-\dfrac{1}{ABv}\exp Bvt\right) + 1\right]} \tag{29}$$

or

$$i = \frac{y'}{\gamma y} = \frac{C \exp Bvt}{A\gamma\left[C + \exp\left(\dfrac{1}{ABv}\exp Bvt\right)\right]} \tag{30}$$

and after substituting A, B, and C, we obtain equation (II.21).

After the variables are substituted, equation (II.20) for the special case $\gamma Q \ll 1$ becomes

$$e^{Bvt} = -A\frac{y'}{y}$$

where

$$A = \frac{\delta Q}{nFSDc^0} = \tau_1 \tag{31}$$

$$y = \gamma Q - \gamma \int_0^t i\,dt; \quad y' = -\gamma i$$

The solution satisfying the initial condition $y = \gamma Q$ at $t = 0$ is expressed by the relationship

$$y = C \exp\left(-\frac{1}{B\upsilon\tau_1}\exp B\upsilon t\right) \tag{32}$$

where

$$C = \gamma Q \exp \frac{1}{B\upsilon\tau_1}$$

Then

$$i = \frac{C}{\gamma\tau_1}\exp\left(-\frac{1}{B\upsilon\tau_1}\exp B\upsilon t\right)\cdot \exp B\upsilon t \tag{33}$$

After the values of the constants are substituted, equation (33) becomes identical to equation (II.22).

The condition for the maximum of this function is

$$\frac{di}{dt} = \frac{C}{\gamma\tau_1}\left[\exp\left(-\frac{1}{B\upsilon\tau_1}\exp B\upsilon t\right)\left(-\frac{1}{\tau_1}\exp B\upsilon t\right)\times\right.$$

$$\left.\times \exp B\upsilon t + \exp\left(-\frac{1}{B\upsilon\tau_1}\exp B\upsilon t\right)\cdot\frac{1}{B\upsilon}\exp B\upsilon t\right] = 0 \tag{34}$$

or

$$B\upsilon\tau_1 = \exp B\upsilon t_{\max}$$

Substituting the value of B and setting $\varphi_1 = \varphi_e$, we easily obtain equation (II.23).

Combining equations (33) and (34), we derive expression (II.24).

Solution of equations (II.16) — (II.18) without the simplifying assumption $c \gg c^0$

Let us combine expressions $(II.16) - (11.18)$ on the basis of $\varphi = \varphi_p + vt$ and consider the case $\gamma Q \ll 1$. Then

$$Bvt = \ln \frac{Q(i\delta + nFSDc^0)}{nFSDc^0 \left(Q - \int_0^t idt\right)} \qquad (35)$$

After making the substitution of variables and abbreviating the result, we obtain the differential equation

$$Bvt = \ln \frac{\tau_1(\lambda - y')}{y} \qquad (36)$$

$$\frac{nFSDc^0}{\delta} = \lambda; \quad Q - \int_0^t idt = y; \quad y' = -i$$

The equation may be represented as

$$\frac{1}{\tau_1} \exp Bvt = \frac{\lambda - y'}{y} \qquad (37)$$

Let us introduce the new functions:

$$g = -\frac{1}{\tau_1} \exp Bvt; \quad E = \exp\left(-\int \frac{1}{\tau_1} \exp Bvt \cdot dt\right);$$

$$u = -\frac{y}{E}, \quad \text{then} \quad \frac{E'}{E} = g; \quad u' = \frac{\lambda}{E}$$

Integrating the last relation and denoting $\exp(Bvt)$ by x, we obtain

$$u = \lambda \int \frac{1}{Bvx} \exp\left(\frac{x}{Bv\tau_1}\right) dx$$

For $\frac{x}{Bv\tau_1} \ll 1$, this expression limited to the first two terms of the series may be represented as:

$$u = \frac{\lambda}{Bv}\left(Bvt + \frac{1}{Bv\tau_1} \exp Bvt + \text{const}\right)$$

which after the reverse substitutions yields

$$y = \frac{\lambda}{Bv}\left(Bvt + \frac{1}{Bv\tau_1}\exp Bvt + \text{const}\right)\exp\left(-\frac{1}{Bv\tau_1}\exp Bvt\right)$$

Using the initial condition $y = Q$ at $t = 0$ and differentiating y with respect to t, we find the value of $i = -y'$

$$i = \frac{Q}{\tau_1}\exp\left(\frac{1}{Bv\tau_1} + Bvt - \frac{1}{Bv\tau_1}\exp Bvt\right) +$$

$$+ \frac{Q}{\tau_1}\cdot\frac{1}{(Bv\tau_1)^2}(\exp Bvt - 1)\exp\left(Bvt - \frac{1}{Bv\tau_1}\exp Bvt\right) +$$

$$+ \frac{Qt}{\tau_1^2}\exp\left(Bvt - \frac{1}{Bv\tau_1}\exp Bvt\right) - \frac{Q}{\tau_1} \qquad (38)$$

It is easy to show that when v is greater than 0.01 V/sec and vt is less than 0.1 (which is usually the case in a reversible process), the ratio of currents obtained with and without consideration of c^0 does not exceed 1.15 to 1.20. At higher values of v this ratio decreases. At the curve maximum, the error due to omitting c^0 is obviously smaller than the value cited above, since t_{max} is less than t (t is the time for complete dissolution of the deposit).

IV. ELECTROCHEMICAL DISSOLUTION OF COMPOUNDS FROM THE SURFACE OF AN INDIFFERENT ELECTRODE

Derivation of formulas (III.15) — (III.18)

Let us apply the Laplace transformation $F(p) = \int_0^t f(t)\exp(-pt)dt$

(p is the transformation coefficient) and transform the equations and boundary conditions (III.12) — (III.14). We obtain

$$D = \frac{d^2\bar{c}_1}{dx^2} = k_2\bar{c}_1 - k_1\bar{c}_2 \qquad (39)$$

$$k_2\bar{c}_1 - k_1\bar{c}_2 = pc_2 - c_2^0 \qquad (40)$$

$$\bar{c}_2(x,0) = c_2^0; \quad \bar{c}_1(x,0) = \frac{c_2^0}{K^1} \qquad (41)$$

where $\bar{c}$ are transformed concentration values.

Using the first shifting theorem, we write the boundary conditions as

$$\frac{\partial \bar{c_1}}{\partial x}\bigg|_{x=0} \exp(-Bvt) = k(\bar{c_1})_{x=0}$$

$$-\frac{\partial \bar{c_1}\,(p + Bv)}{\partial x}\bigg|_{x=0} = k(\bar{c_1})_{x=0} \qquad (42)$$

$$\left(\frac{\partial \bar{c_1}}{\partial x}\right)_{x \to \infty} = 0$$

We solve equation (40) for $\bar{c_2}$ and substitute the resulting value into equation (39)

$$\bar{c_2} = \frac{k_2 \bar{c_1} + c_2^0}{p + k_1}; \qquad D\frac{d^2 \bar{c_1}}{dx^2} = k_2 \bar{c_1} - k_1 \frac{k_2 \bar{c_1} + \bar{c_2^0}}{p + k_1}$$

$$D\frac{\partial^2 \bar{c_1}}{dx^2} - \frac{k_2 \bar{c_1} p}{p + k_1} + \frac{k_1 \bar{c_2^0}}{p + k_1} = 0 \qquad (43)$$

The solution of this equation is

$$\bar{c_1} = \alpha \exp\left(\sqrt{\frac{p}{p + k_1}} \cdot \frac{x}{\mu}\right) + \beta \exp\left(-\sqrt{\frac{p}{p + k_1}} \cdot \frac{x}{\mu}\right) + \frac{c_2^0}{K'p}$$

where $\mu = \sqrt{\dfrac{D}{k_2}}$ is the thickness of the reaction layer.

According to conditions (41) and (42), we obtain

$$\bar{c_1} = \frac{c_2^0}{K'p} + \left[\frac{1}{k}\frac{d\bar{c_1}(p + Bv)}{dx}\bigg|_{x=0} - \frac{c_2^0}{K'p}\right] \exp\left(-\sqrt{\frac{p}{p + k_1}} \cdot \frac{x}{\mu}\right) \qquad (44)$$

Differentiating equation (44) with respect to x and setting $x = 0$, we obtain the following expression for the transformed value of the flux of the substance $Me^{(n+m)+}$ near the electrode surface:

$$\frac{d\bar{c_1}}{dx}\bigg|_{x=0} = \frac{c_2^0}{K'\mu} \cdot \frac{1}{p}\sqrt{\frac{p}{p + k_1}} - \frac{p}{k\mu} \cdot \frac{1}{p}\sqrt{\frac{p}{p + k_1}} \cdot \frac{d\bar{c_1}(p + Bv)}{dx}\bigg|_{x=0} \qquad (45)$$

To restore the functions to the plane of the originals, we use the following formulas, the convolution theorem, and the first shifting theorem:

$$\exp(-\alpha t)\cdot I_0(\alpha t) \longrightarrow \frac{1}{p}\sqrt{\frac{p}{p+2\alpha}}$$

$$I_0(z) = \begin{cases} 1; \quad z \longrightarrow 0 \\ \dfrac{\exp z}{\sqrt{2\pi z}}; \quad z \longrightarrow \infty \end{cases} \qquad (46)$$

where I_0 is a Bessel function for an imaginary argument.

$$p\bar{\varphi}(p)\bar{g}(p) = \frac{d}{dt}\int_0^t \varphi(t-\tau)g(\tau)d\tau$$

Let

$$\bar{\varphi}(p) = \frac{1}{p}\sqrt{\frac{p}{p+k_1}}; \qquad \bar{g}(p) = \left[\frac{d\bar{c}_1(p+Bv)}{dx}\right]_{x=0}$$

Then

$$\frac{\partial c_1(t)}{\partial x}\bigg|_{x=0} = \frac{c_2^0}{K'\mu}\cdot I_0\left(\frac{k_1 t}{2}\right)\exp\left(-\frac{k_1 t}{2}\right) - \frac{1}{K\mu}\times$$

$$\times\frac{d}{dt}\int_0^t \exp\left[-\frac{k_1}{2}(t-\tau)\right]\cdot I_0\left[\frac{k_1(t-\tau)}{2}\right]\times$$

$$\times\exp(-Bv\tau)\cdot\left[\frac{\partial c_1\tau}{\partial x}\right]_{x=0}d\tau$$

$$\frac{d}{dt}\int_0^t \exp\left[-\frac{k_1}{2}(t-\tau)\right]\cdot I_0\left[\frac{k_1(t-\tau)}{2}\right]\cdot\exp(-Bv\tau)\times$$

$$\times\left[\frac{\partial c_1(\tau)}{\partial x}\right]_{x=0}d\tau = \int_0^t\left\{\exp(-Bv\tau)\cdot j(\tau)\cdot\exp\left(\frac{k_1\tau}{2}\right)\times\right.$$

$$(47)$$

$$\times \left[-\frac{k_1}{2} \exp\left(-\frac{k_1}{2} \right) \cdot I_0 \left(\frac{k_1(t-\tau)}{2} \right) + \exp\left(-\frac{k_1 t}{2} \right) \times \right.$$

$$\left. \times I_1 \left(\frac{k_1(t-\tau)}{2} \right) \cdot \frac{k_1}{2} \right] \Big\} \, d\tau + \exp(-Bvt) \cdot j(t) = \qquad (48)$$

$$= \frac{k_1}{2} \int_0^t \exp\left(-Bv\tau \right) \cdot j(\tau) \cdot \exp\left[-\frac{k_1(t-\tau)}{2} \right] \times$$

$$\times \left\{ I_1 \left[\frac{k_1(t-\tau)}{2} \right] - I_0 \left[\frac{k_1(t-\tau)}{2} \right] \right\} d\tau + j(t) \exp\left(-Bvt \right)$$

Here

$$j = \frac{\partial c_1}{\partial x} \bigg|_{x=0}$$

Using equations (47) and (48), we obtain

$$j(t) = \frac{c_2^0}{K'\mu} I_0 \left(\frac{k_1 t}{2} \right) \cdot \exp\left(-\frac{k_1 t}{2} \right) - \frac{1}{k\mu} j(t) \cdot$$

$$\cdot \exp(-Bvt) + \frac{k_1 \exp\left(-\frac{k_1 t}{2} \right)}{2k\mu} \int_0^t \exp\left(-Bv\tau + \frac{k_1 \tau}{2} \right) \times$$

$$\times j(\tau) \cdot \left\{ I_0 \left[\frac{k_1(t-\tau)}{2} \right] - I_1 \left[\frac{k_1(t-\tau)}{2} \right] \right\} d\tau \qquad (49)$$

Hence

$$j(t) = \frac{c_2^0}{K'} \cdot \frac{k \exp\left(-\frac{k_1 t}{2} \right) \cdot I_0 \left(\frac{k_1 t}{2} \right)}{[k\mu + \exp\left(-Bvt \right)]} +$$

$$+ \frac{k_1 \exp\left(-\frac{k_1 t}{2} \right)}{2[k\mu + \exp\left(-Bvt \right)]} \cdot \int_0^t \exp\left(-Bv\tau + \frac{k_1 \tau}{2} \right) \cdot j(\tau) \times \qquad (50)$$

$$\times \left\{ I_0 \left[\frac{k_1(t-\tau)}{2} \right] - I_1 \left[\frac{k_1(t-\tau)}{2} \right] \right\} d\tau$$

Expression (50) multiplied by the product $nFSD$ is the equation for the polarization curve for the stripping of a deposit of a compound from the surface of an indifferent electrode.

Since employing such a complicated integral equation is difficult, we shall consider the limiting case of interest for chemical reactions which are much faster than the electrode potential scanning rate $(k_1 t \gg 1;\ k_1 \gg Bv)$.

We introduce the notation

$$\frac{c_2^0}{K'\mu} = A; \quad k\mu = \varepsilon; \quad \frac{k_1}{2} = k; \quad Bv = \mathrm{v};$$

$$\mathrm{v}\tau = \delta z; \quad \mathrm{v}t = \delta y; \quad \delta = \frac{\mathrm{v}}{k};$$

$$z = k\tau; \quad y = kt; \quad \frac{1}{A\varepsilon}\left[\frac{\partial c_1(t)}{\partial x}\right]_{x=0} = u(y);$$

$$\frac{1}{A\varepsilon}\left[\frac{\partial c_1(\tau)}{\partial x}\right]_{x=0} = u(z)$$

Let us rewrite equation (50) in the form:

$$u(y)\frac{\varepsilon \exp(\delta y)+1}{\exp(\delta y)} = \exp(-y)\cdot I_0(y) +$$

$$+ \int_0^{kt} u(z)\exp(-\delta z)\cdot \exp(z-y)\cdot[I_0(y-z)-I_1(y-z)]dz \tag{51}$$

Expression (51) is easily represented in the following form:

$$u(y)\cdot\frac{\varepsilon \exp \delta y+1}{\exp \delta y} = \exp(-y)\cdot I_0(y) -$$

$$- \int_0^y u(z)\exp(-\delta z)\frac{d}{dy}\{\exp[-(y-z)]I_0(y-z)\}dz \tag{52}$$

Let us take the product $u(z)\exp(-\delta z)$ at the point $z=y$ out from under the integral (below, using the final solution, we shall show that this product varies far more slowly than the expression $\exp(z-y)\cdot[I_0(y-z)-I_1(y-z)]$ and may be assumed constant) and perform the change of variables:

$$y-z=x; \quad dy=dx; \quad dz=-dx.$$

Then expression (52) becomes

$$u(y)\,\frac{\varepsilon\exp\delta y+1}{\exp\delta y}=\exp\,(-y)I_0(y)+u\,(y)\cdot\exp\,(-\delta y)\int_{-y}^{0}\frac{d}{dx}\,[\exp\,(-x)\times$$

$$\times I_0(x)]\,dx \tag{53}$$

or after the integration and transformation:

$$u(y)=\frac{\exp(-y)\cdot I_0(y)\exp(\delta y)}{\varepsilon\exp(\delta y)+\exp(-y)I_0(y)} \tag{54}$$

We introduce functions $f_1(z)$ and $f_2(z)$, which according to expression (54) can be defined as:

$$f_1(z)=u(z)\exp\,(-\delta z)=\frac{\exp\,(-z)\cdot I_0(z)}{\varepsilon\exp\,(\delta z)+\exp\,(-z)\cdot I_0(z)}=$$

$$=\frac{I_0(z)}{\varepsilon\exp\,[z(1+\delta)]+I_0(z)}=\frac{1}{\dfrac{\varepsilon\exp\,(-z)}{I_0(z)}+1} \tag{55}$$

$$f_2(z)=\exp[-\,(y-z)\cdot[I_0(y-z)-I_1(y-z)] \tag{56}$$

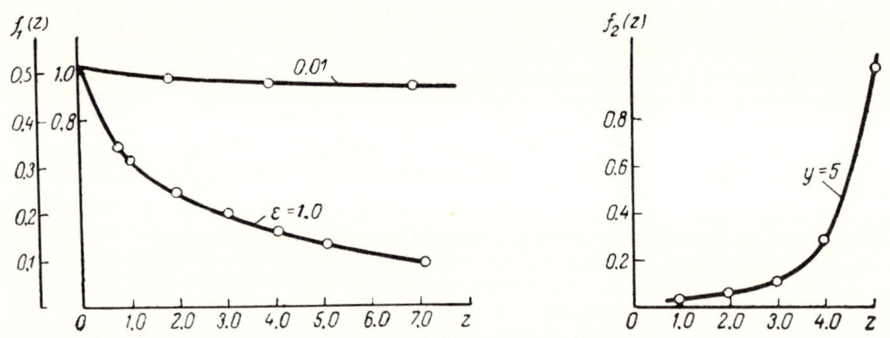

FIGURE 71. Dependence of the functions $f_1(z)$ and $f_2(z)$ on the value of z.

Figure 71 shows that function f_1 varies much more slowly than function f_2 even for values of ε many times greater than its actual value (the change in f_1 decreases with decreasing ε). This justifies the operation used in obtaining expression (53). Recalling that in the present case $k_1t \gg 1$ is large, we may use an asymptotic expansion of the function $I_0(y)$ and represent it as $I_0(y) = \dfrac{\exp(y)}{\sqrt{2\pi y}}$, and then

$$u(y) = \frac{\exp(\delta y)}{\sqrt{2\pi y}\left[\varepsilon \exp(\delta y) + \dfrac{1}{\sqrt{2\pi y}}\right]} \tag{57}$$

We differentiate expression (57) and obtain the condition for the maximum of the function $u(y)$

$$u'(y) = -\frac{\varepsilon\sqrt{\dfrac{\pi}{2y}} - \delta \exp(-\delta y)}{(\varepsilon\sqrt{2\pi y} + \exp \delta y)^2} \tag{58}$$

$$\frac{1}{2}\cdot\frac{\varepsilon}{\delta}\sqrt{2\pi} = \sqrt{y}\,\exp(-\delta y) \tag{59}$$

Returning to the original notation and recalling the equalities $K = K'f_2/f_1$ and $a = c_2^0 f_2$, we easily obtain an equation for the polarization curve for the electrochemical dissolution of a deposit for a linearly varying electrode potential, the relation between the current maximum and the process parameters, and the value of the maximum stripping current

$$i(t) = \frac{nF\cdot SDak\,\exp Bvt}{Kf_1\left[k\mu\,\exp(Bvt)\cdot\sqrt{\pi k_1 t} + 1\right]} \tag{60}$$

$$\frac{k\mu\sqrt{\pi k_1}}{2B\sqrt{v}} = \sqrt{\varphi_{max}}\,\exp(-B\varphi_{max}) \tag{61}$$

$$i_{max} = \frac{nFDSa}{Kf_1\mu}\cdot\frac{\sqrt{v}}{\sqrt{\pi k_1\varphi_{max}}\left(1 + \dfrac{1}{2B\varphi_{max}}\right)} \tag{62}$$

For the limiting cases $a = a_\infty \gamma Q$ (microphase) and $a = a_\infty$ (macrophase), these equations yield expressions (III.15) and (III.17) from equations (III.10) and (III.11).

Taking the logarithm of equation (61), we obtain the expression

$$\varphi_{max} = \frac{1}{2B}\ln 2B\varphi_{max} - \frac{1}{B}\ln k\mu\sqrt{\pi k_1} + \frac{1}{2B}\ln v \tag{63}$$

which is identical to (III.18).

V. ANODIC STRIPPING VOLTAMMETRY OF METALS

TABLE V.1. Concentrating metals on the graphite electrode /172/

Metal ion	Supporting electrolyte	$\varphi_{1/2}$	φ_{el}	φ_{max}
Au^{3+}	$1M$ HCl	$+0.5$	-0.2	$+0.5$
	$1M$ HNO$_3$	$+0.75$	-0.2	$+0.80$
Ag^+	$1M$ KNO$_3$	$+0.05$	-0.2	$+0.10$
Cu^{2+}	$1M$ KNO$_3$	-0.35	-0.6	-0.10
	$0.02\,M$ tartaric acid $+\,0.02M$ NH$_4$Cl $+$ $+$ NaOH, pH$=9$		-0.8	-0.4
	$1M$ KOH $+\,0.1M$ tartaric acid		-1.6	-0.5
	$1M$ KOH $+\,1M$ NaSCN		-1.3	-0.40
Bi^{3+}	$0.1M$ NH$_4$Cl	-0.2	-0.5	-0.15
	$0.1M$ KCl $+\,0.1M$ HCl	-0.2	-0.4	-0.10
	$0.1M$ NH$_4$Cl	-0.2	-0.4	-0.1
Pb^{2+}	$0.1M$ HCl	-0.6	-1.0	-0.55
	$0.1M$ NH$_4$Cl	-0.6	-0.8	-0.55
	20% citric acid	-0.7	-1.0	-0.50
Sb^{3+}	$1M$ NH$_4$CH$_3$COO	-0.3	-1.0	-0.30
	$1M$ HCl	-0.25	-0.5	-0.20
Sn^{2+}	$1M$ NH$_4$Cl	-0.75	-1.2	-0.50
	$1M$ HCl	-0.6	-1.2	-0.50
	20% citric acid		-1.2	-0.25
Cd^{2+}	$0.1M$ HCl	-0.8	-1.2	-0.75
	$1M$ KCl		-1.2	-0.70
Hg^{2+}	$1M$ KSCN	-0.25	-0.6	-0.1
	$0.1M$ KNO$_3$	-0.1	-0.4	$+0.1$
In^{3+}	$1M$ NH$_4$SCN	-0.9	-1.4	-0.7
	$0.1M$ KCl	-0.9	-1.3	-0.7
Tl^+	$0.2M$ (NH$_4$)$_2$SO$_4$ $+$ NH$_4$OH, pH$=8$	-0.95	-1.4	-0.8
	$0.1M$ NaSCN		-1.2	-0.4
Ni^{2+}	$0.1M$ NH$_4$OH $+\,0.1M$ NH$_4$Cl	-1.1	-1.2	-0.5
	$1M$ KNO$_3$ $+\,0.001M$ HNO$_3$		-1.2	-0.1
	$0.1M$ NaSCN		-1.2	-0.5
Co^{2+}	$0.2M$ NH$_4$OH		-1.2	-0.4
	$0.1M$ K$_2$SO$_4$		-1.4	-0.1
	$0.1\,M$ tartaric acid $+\,1\,M$ KOH		-1.6	-0.6
	$1M$ KOH $+\,0.1\,M$ NaSCN		-1.2	-0.7
Fe^{3+}	$1M$ KOH $+\,0.01\,M$ tartaric acid		-1.4	-0.9
	$0.05\,M$ NaC$_7$H$_5$O$_6$S, pH$=5$		-1.6	-0.6

N o t e . Approximate values are listed of $\varphi_{1/2}$ for metal ion concentrations of $1\cdot10^{-5}$ to $1\cdot10^{-4}$ g-ion/liter and of φ_{max} for $1\cdot10^{-7}$ to $1\cdot10^{-6}$ g-ion/liter at a potential scanning rate of 0.017 V/sec.

TABLE V.2. Application of anodic stripping voltammetry of metals in analysis

Substance analyzed	Element deter-mined	Type of electrode	Supporting electrolyte	Sensitivity*		Refer-ence
				g-ion/liter	wt.%	
Salts of alkali metals and alkaline earths	Ag	Graphite (type I)	Solution of salt being analyzed		$1 \cdot 10^{-5}$	69
$Pb(NO_3)_2$, $Pb(CH_3COO)_2$	Ag	Same	Same		$1 \cdot 10^{-5}$	72
Thiourea**	Ag	"	0.5 N $NaCH_3COO$ + HNO_3, pH = 4		$5 \cdot 10^{-5}$	72
KNO_3	Ag	Graphite (type II)	0.2 M KNO_3	$2.5 \cdot 10^{-9}$		70
KNO_3	Ni	Pt, Au	0.1 M KNO_3 + 0.01 M KSCN	$5 \cdot 10^{-8}$		220
KNO_3	Ag	Graphite, paraffin-impregnated	Solution of salt being analyzed	$4 \cdot 10^{-9}$		174
	Ag	Same	Same	$1.5 \cdot 10^{-10}$		82
HCl	Au	"	0.1 M HCl			185
HNO_3	Cu	Mercury film electrode	0.1 M HNO_3			200
Li_2SO_4	Hg	Graphite (type II)	1.8 M KSCN + 0.1 M KSCN			200
		Graphite, wax-impregnated	0.1 M KSCN			75
Fused fluorides	Ni	Platinum	Fused fluorides			200
Cd	Sb	Graphite (type I)	$CdCl_2$ + 2 M HCl + + $1 \cdot 10^{-4}$ M $Hg(NO_3)_2$	$2 \cdot 10^{-7}$	$5 \cdot 10^{-5}$	
Pb	Sb	Same	6 M HCl + 0.06 M $Pb(NO_3)_2$ + + $1 \cdot 10^{-4}$ M $Hg(NO_3)_2$	$3 \cdot 10^{-8}$	$5 \cdot 10^{-5}$	
KSCN	Hg	Graphite, wax-impregnated	0.1 M KSCN	$4 \cdot 10^{-9}$		75
CdS, ZnS	Cu	Graphite (type I)	$CdCl_2$, $ZnCl_2$	$1 \cdot 10^{-7}$	$3 \cdot 10^{-6}$	
$(NH_4)_2MoO_4$*	Co	Same	1.25 N NaSCN + 2.5 N KOH	$2 \cdot 10^{-7}$	$1 \cdot 10^{-5}$	

TABLE V.2. continued

Substance analyzed	Element determined	Type of electrode	Supporting electrolyte	Sensitivity*		Reference
				g-ion/liter	wt.%	
$(NH_4)_2MoO_4$**	Fe	Graphite (type I)	1.5 N KOH + 0.005 M tartaric acid + 0.001 M Na_2S	$2\cdot10^{-7}$	$1\cdot10^{-5}$	
$(NH_4)_2MoO_4$**	Cu	Same	1 M KOH + 0.04 M tartaric acid + 0.1 M NH_4Cl	$2\cdot10^{-7}$	$1\cdot10^{-5}$	
Citric and tartaric acids	Sn	"	1 M citric (tartaric) acid +	$2\cdot10^{-7}$	$1\cdot10^{-5}$	
	Sb	"	+ 8 M HCl + + $1\cdot10^{-4}$ M $Hg(NO_3)_2$			
	Pb	"	1 M citric (tartaric) acid + + $1\cdot10^{-4}$ M $Hg(NO_3)_2$	$1\cdot10^{-7}$	$1\cdot10^{-5}$	
	Fe	"	1 M potassium citrate(tartrate) + + KOH + $5\cdot10^{-3}$ M Na_2S, pH $=10$	$2\cdot10^{-7}$	$1\cdot10^{-5}$	
HNO_3	Cu, Pb	"	0.1 N HNO_3 + + $1\cdot10^{-5}$ M $Hg(NO_3)_2$	$1\cdot10^{-8}$	$1.5\cdot10^{-6}$	81
KNO_3	Tl Pb	Graphite, paraffin-impregnated	Solution of salt being analyzed	$5.5\cdot10^{-8}$		82
KCl	Cd	Same	0.0002 M KCl	$6\cdot10^{-11}$		221
$(NH_4)_2C_2O_4$	Hg, Ag	Graphite (type I)	1 M KSCN			
$K_2C_2O_4$	Sum of Cu and Bi	Graphite (type I)	1.5% potassum citrate + salt being analyzed		$(2-4)\cdot10^{-6}$	
	Pb, Cd	Same	$1.2\cdot10^{-4}$ M $Hg(NO_3)_2$ + salt being analyzed		$(2-4)\cdot10^{-6}$	71
HCl	Pb, Cd Sum of Cu and Bi	"	0.5 M HCl + + $1\cdot10^{-4}$ M $Hg(NO_3)_2$		$1\cdot10^{-7}$	

$Er_2(MoO_4)_3$** $Y_2(MoO_4)_3$ $Nd_2(MoO_4)_3$	Fe	"	1 M KOH + 0.05 M tartaric acid		10^{-4}	84
Cd, Pb	Ag	"	$Me(NO_3)_2$ + HNO_3, pH = 1	$2 \cdot 10^{-7}$	$2 \cdot 10^{-5}$	
Sn	Ag	"	$Sn(SO_4)_2$ + 5 M H_2SO_4	$5 \cdot 10^{-7}$	$1 \cdot 10^{-4}$	
Sb	Au	Graphite (type II)	1 M HCl	$5 \cdot 10^{-8}$	$5 \cdot 10^{-6}$	67
Plant ash**	Au	Same	Same			
Pb	Ag	Graphite (type I)	$Pb(NO_3)_2$ + 0.1 M KSCN		$1 \cdot 10^{-7}$	
As and GaAs	Ag	Graphite (type II)	$GaBr_3$ + 0.01 M NH_4OH + 0.001 M NH_4NO_3		$2 \cdot 10^{-6}$	68
Bronze and brass	Pb	Graphite (type I)	0.1 M CH_3COOH + 0.1 M CH_3COONa		$1 \cdot 10^{-4}$	206
Cu	Pb	Same	1 M KCl, pH = 6		$1 \cdot 10^{-5}$	
	Cd	"	1 M KCl, pH = 6		$5 \cdot 10^{-6}$	206
	Sb, Bi	"	1 M KCl + 4% citric acid		$5 \cdot 10^{-6}$	222
Zinc and zinc alloys	Pb, Cu	"	0.1 M CH_3COOH + 0.1 M $NaCH_3COO$		$1 \cdot 10^{-4}$	
	In, Hg	"	2 M KSCN		$1 \cdot 10^{-5}$	223
	Tl	"	1 M NH_4Cl + 0.25 M EDTA		$1 \cdot 10^{-5}$	
Pb	Ag	"	0.05 M KSCN		$1 \cdot 10^{-4}$	224
	Sb, Bi	"	8 N HCl + 4% citric acid		$1 \cdot 10^{-5}$	

* The column "Sensitivity" lists concentrations which may be determined with an error not exceeding 20%.

** The preparation is calcined and then dissolved in the supporting electrolyte.

VI. STRIPPING VOLTAMMETRY OF VARIABLE-VALENCE IONS

TABLE VI.1. Concentrating variable-valence ions as compounds with inorganic reagents

Electro-active element	Electrode reaction	Solution composition	$\varphi_{1/2}$	φ_{el}	φ_{max}	c_{min} g-ion/ /liter	Refer-ence
Cr	$CrO_4^- + 4H_2O + 3e \rightleftharpoons$ $\rightleftharpoons Cr(OH)_3 + 5OH^-$	$0.1\,M\,NaOH + CrO_4^{2-}$	-0.8	-1.0	$+0.20$		
		$0.05\,M\,Na_2B_4O_7 + CrO_4^{2-}$. pH$=$9.2	-0.7	-1.0	$+0.6$	$2\cdot10^{-7}$	
		$C_6H_8O_7 + NaOH + CrO_4^{2-}$. pH$=$5.2	-0.35	-0.5	$+0.7$	$1\cdot10^{-6}$	
		$KH_2PO_4 + NaOH + CrO_4^{2-}$. pH$=$8.0	-0.35	-0.7	$+0.7$	$1\cdot10^{-7}$	99
		$0.4\,M\,NH_4Cl + 0.1\,M\,NH_4OH + CrO_4^{2-}$	-0.45	-0.7	$+0.6$	$4\cdot10^{-8}$	
		$0.1\,M\,HCH_3COO + 0.1\,M\,NaCH_3COO + + CrO_4^{2-}$	-0.1	-0.5	$+0.9$	$2\cdot10^{-6}$	
		$0.01\,M\,H_2SO_4 + CrO_4^{2-}$	$+0.15$	-0.3	$+1.20$	$1\cdot10^{-5}$	
	$Cr^{3+} + 8OH^- + Me^{2+} \rightleftharpoons$ $\rightleftharpoons MeCrO_4 + 4H_2O + 3e$	$Ba(OH)_2 + Cr^{3+}$	$+0.2$	$+0.3$	-0.6	$6\cdot10^{-6}$	92
	Me—Ba, Pb	$Pb(CH_3COO)_2 + Cr^{3+} + + 0.1\,M\,NaCH_3COO$	$+0.8$	$+1.0$	-0.05	$1\cdot10^{-4}$	
Ce	$Ce^{3+} + 4H_2O \rightleftharpoons$ $\rightleftharpoons Ce(OH)_4 + 4H^+ + e$	$0.1\,M\,HCH_3COO + + 0.1\,M\,NaCH_3COO + Ce^{3+}$	$+0.7$	$+1.0$	$+0.3$	$1\cdot10^{-6}$	98

Tl	$Tl^+ + 3OH^- \rightleftharpoons Tl(OH)_3 + 2e$	$0.35\,M\,(NH_4)_2SO_4 + H_2SO_4 + Tl^+$	$+1.25$	$+1.5$	$+0.5$	$1 \cdot 10^{-5}$	
		pH=4					
		$0.35\,M\,(NH_4)_2SO_4 + Tl^+$	$+1.15$	$+1.4$	$+0.35$	$2 \cdot 10^{-6}$	100
		$0.35\,M\,(NH_4)_2SO_4 + NH_4OH + Tl^+$					
		pH=7	$+1.0$	$+1.2$	$+0.3$	$5 \cdot 10^{-7}$	
		pH=8	$+0.8$	$+1.1$	$+0.2$	$2 \cdot 10^{-6}$	
		pH=9	$+0.7$	$+0.9$	$+0.15$	$3 \cdot 10^{-6}$	
		pH=10	$+0.55$	$+0.8$	$+0.1$	$9 \cdot 10^{-5}$	
Mn	$Mn^{2+} + 4OH^- \rightleftharpoons Mn(OH)_4 + 2e$	$0.1\,M\,HNO_3 + Mn^{2+}$	$+1.2$	$+1.3$	$+0.9$		
		$2\,M\,(NH_4)_2SO_4 + H_2SO_4 + Mn^{2+}$					
		pH=3	$+0.9$	$+1.1$	$+0.7$		
		pH=4	$+0.8$	$+1.0$	$+0.65$		
		pH=5	$+0.75$	$+0.9$	$+0.25$	$5 \cdot 10^{-7}$	96, 186
		$2\,M\,(NH_4)_2SO_4 + Mn^{2+}$					
		pH=7	$+0.5$	$+0.7$	$+0.15$		
		pH=8	$+0.25$	$+0.5$	$+0.10$		
		pH=9	$+0.15$	$+0.4$	$+0.05$		
Fe	$Fe^{2+} + 3OH^- \rightleftharpoons Fe(OH)_3 + e$	$H_3BO_3 + NaOH + Fe^{2+}$					
		pH=8		-0.05	-0.5	$1 \cdot 10^{-7}$	101
	$Fe^{3+} + 2OH^- + e \rightleftharpoons Fe(OH)_2$	0.5% citric acid + NaOH + Fe^{3+}	-0.85	-1.0	-0.7	$2 \cdot 10^{-6}$	
		pH=10					

TABLE VI.2. Concentrating variable-valence ions as compounds with organic reagents

Electro-active element	Electrode reaction	Solution composition	$\varphi_{1/2}$	φ_{el}	φ_{max}	Sensitivity g-ion/liter	Reference
Co	$Co^{2+} + 3RH = CoR_3 + 3H^+ + e$	$0.4\,M\,NH_4OH + 0.05\,M\,NH_4Cl + {}$ $+ 1.5\cdot10^{-4}\%RH + Co^{2+}$	-0.60	-0.5	-0.7	$1\cdot10^{-8}$	104
Sb	$[SbCl_6]^{3-} + R^+ \rightleftarrows R[SbCl_6] + 2e$	$1.5\,M\,KCl + 0.5\,M\,H_2SO_4 + {}$ $+ 4\cdot10^{-4}\,M\,R + Sb^{3+}$		$+0.8$	$+0,35$		105
I	$2I^- + Cl^{--} + R^+ \rightleftarrows R[I_2Cl] + 2e$	$0.1\,M\,H_2SO_4 + 5\cdot10^{-5}\,M\,R + {}$ $+ 0.1\,M\,KCl + I^-$	$+0.55$	$+0.8$	$+0.35$	$5\cdot10^{-9}$	106
	$2I^- + SCN^- + (C_2H_5)_4N^+ \rightleftarrows$ $\rightleftarrows (C_2H_5)_4N[I_2SCN] + 2e$	$0.1\,M\,H_2SO_4 + 5\cdot10^{-5}\,M\,R + {}$ $+ 0.1\,M\,KSCN + I^-$		$+0.55$	$+0.4$		
Ni	$Ni(DH)_n + 2OH \longrightarrow NiODH + e$	$0.03\,M\,KOH + 1\cdot10^{-6}\,M\,H_2D + Ni^{2+}$		$+0.8$	$+0.4$	$2\cdot10^{-9}$	107

Note. RH — nitrosonaphthol, R — rhodamine S, H_2D — dimethylglyoxime.

TABLE VI.3. Application of stripping voltammetry of variable-valence ions in analysis

Substance analyzed	Element deter-mined	Supporting electrolyte	Sensitivity		Refer-ence
			g-ion/ /liter	wt.%	
Compounds of rare earths	Ce	$MeCl_3 + 0.1\,N\ NaCH_3COO + 0.1\,N.\ CH_3COOH^1$		$4 \cdot 10^{-5}$	98
$PbMoO_4^*$, $(NH_4)_2MoO_4^{**}$, Na_2WO_4	Cr	$2.5\,M\,NaOH + 0.2\,M\,Na_2H_2EDTA + HNO_3;\ pH{=}8^3$	$1 \cdot 10^{-6}\,M$	$1 \cdot 10^{-4}$	99
$CuSO_4$	Tl	$0.5\,M\,CuSO_4 + 0.5\,M\,(NH_4)_2SO_4 + NH_4OH;\ pH{=}8$	$2 \cdot 10^{-6}$	$7 \cdot 10^{-4}$	100
$CdSO_4$	Tl	$2\,M\,CdSO_4 + 2.5\,M\,(NH_4)_2SO_4 + NH_4OH;\ pH{=}8.5$	$2 \cdot 10^{-6}$	$1 \cdot 10^{-4}$	100
$CdSO_4$	Mn	$2\,M\,CdSO_4 + NH_4OH;\ pH{=}5$	$5 \cdot 10^{-7}$	$2.5 \cdot 10^{-5}$	96
$HClO_4$	Mn	$0.1\,M\,HClO_4$	$1 \cdot 10^{-7}$		95
Alkali metal hydroxides[2]	Mn	$MeOH + 1.5\,N.\ NH_4Cl$	10^{-9}	10^{-7}	112
$NaCH_3COO^2$	Mn	$0.2\,N.\ NaCH_3COO + CH_3COOH;\ pH{=}5$	$5 \cdot 10^{-9}$		122
KCl	Sb	$2.5\,M\,KCl + 4 \cdot 10^{-4}\,M\,R + 0.1\,M\,H_2SO$	$2 \cdot 10^{-8}$	$3 \cdot 10^{-6}$	105
KCl	I	$1\,M\,KCl + 6 \cdot 10^{-5}\,M\ R + 0.1\,M\,H_2SO_1 + {}$ $+ 5 \cdot 10^{-3}\,M\,N_2H_4$	$6 \cdot 10^{-9}$	$2 \cdot 10^{-6}$	106
Seawater	I	$Seawater + 5 \cdot 10^{-3}\,M\,N_2H_4 + 6 \cdot 10^{-5}\,M\,R + {}$ $+ 0.2\,M\,HCl$		$1 \cdot 10^{-7}$	106

TABLE VI.3. continued

Substance analyzed	Element deter-mined	Supporting electrolyte	Sensitivity		Refer-ence
			g-ion//liter	wt.%	
$Al(NO_3)_3$	Ni	$0.15\,M\,Al(NO_3)_3 + 0.5\,M\,KOH + 5\cdot10^{-6}\,M\,H_2D^7$	$5\cdot10^{-9}$	$3\cdot10^{-6}$	
$CaCl_2$	Co	$1\,M\,CaCl_2 + 0.1\,M\,NH_4Cl + NH_4OH + {} + 2\cdot10^{-4}\,M\,RH.\ pH=8$	$2\cdot10^{-8}$	$1\cdot10^{-6}$	
$(Er,\ Y,\ Nd)_2(MoO_4)_3$	Co	$0.25\,M$ citric acid $+ 0.4\,M\,NH_4OH + {} + 0.05\,M\,NH_4Cl + NaOH;\ pH=8$	$1\cdot10^{-8}$	$3\cdot10^{-5}$	
Na_2WO_4*	Ni	$0.2\,M\,KOH + 1\cdot10^{-5}\,M\,H_2D$	$1\cdot10^{-7}$	$1\cdot10^{-5}$	
Citric acid	Co	$0.5\,M$ sodium citrate $+ NaOH;\ pH=10\div11$	$1\cdot10^{-7}$	$1\cdot10^{-5}$	
Tartaric acid	Co	$0.5\,M$ sodium tartrate $NaOH;\ pH=9$	$1\cdot10^{-7}$	$1\cdot10^{-5}$	
Citric and tartaric acids†	Ni	$0.1\,M\,NaOH + 1\cdot10^{-5}\,M\,H_2D$	$2\cdot10^{-7}$	$1\cdot10^{-5}$	
Bronze and brass‡	Sb	$1\,M\,HCl + 1\cdot10^{-3}\,M\,R$		$5\cdot10^{-4}$	
Zn	Ce	$0.1\,M\,CH_3COOH + 0.1\,M\,CH_3COONa$		$1\cdot10^{-4}$	223

Notes: 1. Me^{n+} is the metal ion from the compound analyzed, H_4EDTA is ethylenediaminetetraacetic acid, R is rhodamine S, and H_2D is dimethylglyoxime.

2. In the analysis of $HClO_4$, alkali metal hydroxides, and NaO_2CCH_3, a platinum electrode is employed, in all other cases, a graphite (type I) electrode.

3. The column "Sensitivity" lists concentrations which may be obtained with an error not exceeding 20%

* Dissolved in an NaOH or KOH solution.

** Dissolved in a 10% NaOH solution and neutralized with HNO_3.

† Roasted and then dissolved in the supporting electrolyte.

‡ Dissolved in nitric acid and evaporated with hydrochloric acid.

VII. CATHODIC STRIPPING VOLTAMMETRY OF ANIONS

TABLE VII.1. Concentrating anions as mercury salts

Anion	Supporting electrolyte	t, °C	φ_{el}	φ_{max}	c^0_{min}, g-ion/ /liter
Cl^-	0.1 N NaNO$_3$	20	+0.35	+0.15	$5 \cdot 10^{-6}$
	''	2			$1 \cdot 10^{-6}$
Br^-	''	20	+0.25	0.0	$1 \cdot 10^{-6}$
		2			$5 \cdot 10^{-7}$
I^-	''	20	+0.1	—0.3	$5 \cdot 10^{-8}$
S^{2-}	1 N NaOH	20	—0.5	—0.9	$5 \cdot 10^{-8}$
CrO_4^{2-}	0.1 N NaNO$_3$	20		—0.35	$3 \cdot 10^{-9}$
WO_4^{2-}	''	20	+0.4	+0.25	$4 \cdot 10^{-7}$
MoO_4^{2-}	''	20	+0.4	+0.20	$1 \cdot 10^{-6}$
VO_3^-	''	20	+0.4	+0.25	$1 \cdot 10^{-6}$
SO_4^{2-}*	''	20	—0.5	—0.9	$5 \cdot 10^{-8}$
Oxalate ions	1 N KNO$_3$	20	+0.35	+0.05	$1 \cdot 10^{-6}$
Succinate ions	''	20	+0.40	0.0	$1 \cdot 10^{-4}$
Dithizonate ions	''	20	+0.1	—0.35	$1 \cdot 10^{-5}$
Diethyldithiophosphate ions	''	20	0.0	—0.3	$1 \cdot 10^{-6}$

* Can be determined only after reduction to S^{2-}, for example, by a solution of titanium (III) in phosphoric acid.

TABLE VII.2. Application of cathodic stripping voltammetry of anions in analysis

Substance analyzed	Anion determined	Stripping conditions	Type of electrode	Supporting electrolyte	Sensitivity		Reference
					g-ion/liter	wt.%	
K_2CO_3, Na_2CO_3	SO_4^{2-} *	Linearly varying potential	HMDE (Pt)	5 M KOH	$5 \cdot 10^{-8}$	$5 \cdot 10^{-6}$	150
KNO_3	Cl^-	Same	"	0.08 M KNO_3 + 0.002 M HNO_3	$5 \cdot 10^{-6}$		145
$UOSO_4$	Cl^-	"	Mercury micro-electrode		$5 \cdot 10^{-6}$		148
$CaWO_4$, $SrWO_4$**, $CaMoO_4$, $SrMoO_4$	Cl^-	Same	HMDE (Pt)	1.2 M tartaric acid + + 0.5 M HNO_3	$5 \cdot 10^{-6}$	$5 \cdot 10^{-4}$	153
$BaCo_3$, $SrCO_3$, $CaCO_3$	I^-	Linearly varying potential Oscillographic polarography	"	$Me(NO_3)_2$	$5 \cdot 10^{-8}$	$5 \cdot 10^{-6}$	155
$NaNO_3$	I^-	Linearly varying potential	HMDE	$NaNO_3$	$1 \cdot 10^{-6}$		147
$CdSO_4$	$CrSO_4^{2-}$	Same	HMDE (Pt)	0.5 M $CdSO_4$ + HNO_3; pH = 4	$2 \cdot 10^{-9}$	$2 \cdot 10^{-7}$	157
H_2O	S^{2-}	"	HMDE	0.1 N NaOH	$1 \cdot 10^{-7}$	$3 \cdot 10^{-7}$	156
$NaCH_3COO$ + + HCH_3COO	I^-	"	"	0.1 M $NaCH_3COO$ + + 0.1 M CH_3COOH	$4 \cdot 10^{-8}$		154
	Cl^-	"	"	0.1 M $NaCH_3COO$ + 80% C_2H_5OH	$4 \cdot 10^{-8}$		
$Al(NO_3)_3$	Cl^-	"	HMDE (Pt)	0.5 M $Al(NO_3)_3$ + 0.8 M HNO_3 + + 80% C_2H_5OH; $t° = 2°C$	$7 \cdot 10^{-7}$	$1 \cdot 10^{-5}$	
$Al(NO_3)_3$†	SO_4^{2-} *	"	"	5 M KOH + 0.5 M N_2H_4	$5 \cdot 10^{-8}$	$1 \cdot 10^{-5}$	
KIO_3	I^-	Oscillographic polarography	"	0.4 M KIO_3	$8 \cdot 10^{-8}$	$1 \cdot 10^{-5}$	
	S^2‡	Linearly varying potential	Graphite (ED-6)	0.4 M KIO_3	$8 \cdot 10^{-8}$	$1 \cdot 10^{-5}$	
Tl	Se^{4+}	Vector polarography	HMDE	(0.4—1.3) M HCl + 4—6% HNO_3	$5 \cdot 10^{-9}$	$2 \cdot 10^{-5}$	207
Ores	Se^{4+}	Same		0.25 M H_2SO_4			207
Se	Tl^+	"	"	0.2 M HI	$2 \cdot 10^{-9}$	$1 \cdot 10^{-6}$	208

* SO_4^{2-} is reduced to S^{2-} by titanium (III) phosphate, and the H_2S is absorbed by 5M KOH.

** The preparations are reduced in a zinc amalgam in a tartrate medium.

† The preparation is calcined.

‡ An indirect method from the excess unreacted mercury in the solution.

HMDE — hanging mercury drop electrode.

BIBLIOGRAPHY

1. Stromberg, A.G. and E.A. Stromberg. — Zavodskaya Laboratoriya **27** (1961), 3.
2. Stromberg, A.G. and E.A. Zakharova. — Zavodskaya Laboratoriya 30 (1964), 3, 261.
3. Brainina, Kh.Z., N.K. Kiva, and V.B. Belyavskaya. — Elektrokhimiya 1 (1965), 3, 311.
4. Vetter, K. Electrochemical Kinetics. — New York, Academic Press, 1967.
5. Frumkin, A.N., V.S. Bagotskii, Z.A. Iofa, and B.N. Kabanov. Electrode Process Kinetics. — Moscow State Univ., 1952. (Russian)
6. Mairanovskii, S.G. Catalytic and Kinetic Waves in Polarography. — "Nauka," 1966. (Russian)
7. Kryukova, T.A., S.I. Sinyakova, and T.V. Aref'eva. Polarographic Analysis. — Goskhimizdat, 1959. (Russian)
8. Delahay, P. New Instrumental Methods in Electrochemistry. — New York, Interscience Publishers, 1966.
9. Koutecký, J. — Colln. Czech. Chem. Commun. 18 (1953), 183.
10. Koutecký, J. — Colln. Czech. Chem. Commun. 18 (1953), 311.
11. Koutecký, J. — Colln. Czech. Chem. Commun. 18 (1953), 597.
12. Koutecký, J. — Colln. Czech. Chem. Commun. 19 (1954), 1045.
13. Koutecký, J. — Colln. Czech. Chem. Commun. 19 (1954), 857.
14. Koutecký, J. — Colln. Czech. Chem. Commun. 19 (1954), 1093.
15. Koutecký, J. — Colln. Czech. Chem. Commun. 20 (1955), 116.
16. Koutecký, J. — Colln. Czech. Chem. Commun. 21 (1956), 652.
17. Koutecký, J. — Colln. Czech. Chem. Commun. 21 (1956), 1056.
18. Koutecký, J. — Colln. Czech. Chem. Commun. 22 (1957), 160.
19. Tafel, J. — Z. phys. Chem. 50 (1905), 641.
20. Shain, I. and W.G. Stevens. — Rev. Polarogr. 14 (1967), 232.
21. Delahay, P. Double Layer and Electrode Kinetics. — New York, Interscience Publishers, 1966.
22. Carslaw, H.S. and J.C. Jaeger. Conduction of Heat in Solids (2nd edition). — Oxford Univ. Press, 1959.
23. Levich, V.G. Physicochemical Hydrodynamics. — Fizmatgiz, 1959. (Russian) [English translation: New Jersey, Prentic Hall, 1962]
24. Brdička, R. and K. Wisner. — Naturwissenschaften 31 (1943), 247.
25. Wisner, K. — Z. Elektrochem. 49 (1943), 164.
26. Wisner, K. — Chemické Listy 41 (1947), 6.
27. Koutecký, J. and R. Brdička. — Colln. Czech. Chem. Commun. 12 (1947), 337.
28. Budevki, E. — In: Izvestiya na Bŭlgarskata Akademiya na Naukite, Dep. phys. — math. sci., phys. ser. 3 (1952—1954), 43.
29. Kern, D. — J. Am. Chem. Soc. 75 (1953), 2473.
30. Hanuš, V. — Chemické Zvesti 8 (1954), 702.
31. Kabanov, B.N. Electrochemistry of Metals and Adsorption. — "Nauka," 1966. (Russian)
32. Khor, T.P. — In: New Problems in Modern Electrochemistry. "Mir," (1962). (Russian)

33. B o c k r i s , J. and A. D a m j a n o v i c . — In: Bockris, J. O'M. and B.E. Conway (eds.).
    Modern Aspects of Electrochemistry, No.3. London, Butterworths, 1964.
34. H a i s s i n s k y , M. Nuclear Chemistry and its Applications. — Reading, Addison-Wesley,
    1964.
35. H a i s s i n s k y , M. — J. Chim. phys. 32 (1935), 116.
36. H a i s s i n s k y , M. — J. Chim. phys. 43 (1946), 21.
37. R o g e r s , L.B. and A.E. S t e h n e y . — Trans. Electrochem. Soc. 95 (1949), 25.
38. R o g e r s , L.B., D.R. K r a u s e , and J.C. G r i e s s . — Trans. Electrochem. Soc. 95 (1949), 2,
39. G r i e s s , J.C. Jr. and L.B. R o g e r s . — Trans. Electrochem. Soc. 95 (1949), 129.
40. D e G e i s o , R.C. and L.B. R o g e r s . — J. Electrochem. Soc. 106 (1959), 433.
41. R o g e r s , L.B. — Record of Chem. Progr. 16 (1955), 197.
42. G r i e s s , J.C. Jr., J.T. B y r n e , and L.B. R o g e r s . — J. Electrochem. Soc. 98 (1951), 11, 44
43. B y r n e , J.T., L.B. R o g e r s , and J.C. G r i e s s Jr. — J. Electrochem. Soc. 98 (1951), 11, 45
44. B y r n e , J.T. and L.B. R o g e r s . — J. Electrochem. Soc. 98 (1951), 11, 457.
45. Z i v , D.M. and V.A. J s h i n a . — Radiokhimiya 1 (1959), 2, 185.
46. Z i v , D.M., V.A. I s h i n a , and V.S. Z i v . — Radiokhimiya 1 (1959), 4, 488.
47. Z i v , D.M. and G.S. S i n i t s y n a . — In: Trudy Radievyi Inst. im. V.G. Khlopina Akad.
    Nauk SSSR 8 (1958), 127, 138.
48. N i k o l ' s k i i , B.N., G.S. S i n i t s y n a , and D.M. Z i v . — 8 (1958), 141.
49. L e w i s , G.N. and M. R a n d e l l . Thermodynamics and the Free Energy of Chemical
    Substances. — New York, 1923.
50. H a i s s i n s k y , M. and A. C o c h e . — J. Chem. Soc., Suppl. 2 (1949), 397.
51. D a n o n , J. and M. H a i s s i n s k y . — J. Chim. phys. 47 (1950), 951.
52. C o c h e , A. — Compt. rend. 225 (1947), 936.
53. H a i s s i n s k y , M., H. F a r a g g i , A. C o c h e , and P. A v i g n o n . — Phys. Rev. 75 (1949),
    1963.
54. K a i s h e v , R. and B. M u t a f c h i e v . — In: Referaty Dokladov 14 Soveshchaniya po Elektro-
    khimicheskoi Termodinamike i Kinetike, 19—24 August, 1963, p.34.
55. B r a i n i n a , Kh.Z. and E.M. R o i z e n b l a t . — Elektrokhimiya 1 (1965), 4, 403.
56. B r a i n i n a , Kh.Z. — Zhurn. Analit. Khim. 18 (1963), 10, 1169.
57. N i c h o l s o n , M.M. — J. Am. Chem. Soc. 79 (1957), 7.
58. B r a i n i n a , Kh.Z., E.M. R o i z e n b l a t , and V.B. B e l y a v s k a y a . — Zavodskaya
    Laboratoriya 28 (1962), 9, 1047.
59. B r a i n i n a , Kh.Z. and G.V. Y a r u n i n a . — Elektrokhimiya 2 (1966), 7, 781.
60. B r a i n i n a , Kh.Z. — Elektrokhimiya 2 (1966), 8, 901.
61. B r a i n i n a , Kh.Z. — Elektrokhimiya 2 (1966), 9, 1006.
62. B r a i n i n a , Kh.Z. and V.B. B e l y a v s k a y a . — Elektrokhimiya 2 (1966), 10, 1158.
63. B r a i n i n a , Kh.Z. and V.B. B e l y a v s k a y a . — Elektrokhimiya 3 (1967), 10, 1283.
64. S a t h y a n a r a y n a , S. — J. Elektroanal. Chem. 7 (1964), 403.
65. D e l a h a y , P. and C. M a t t a x . — J. Am. Chem. Soc. 76 (1954), 874.
66. M o r i n a g a , K. — Bull. Chem. Soc. Japan 29 (1956), 793.
67. S i n y a k o v a , S.I. and L.S. C h u l k i n a . — Zhurn. Analit. Khim. 23 (1968), 6, 841.
68. C h u l k i n a , L.S. and S.I. S i n y a k o v a . — Zhurn. Analit. Khim. 24 (1969), 2, 247.
69. B r a i n i n a , Kh.Z., T.A. R y g a i l o , and V.B. B e l y a v s k a y a . In: Metody Analiza
    Khimicheskikh Reaktivov i Preparatov, No. 5—6 (1963), 124.
70. M o n e n , H., H. S p e c k e r , and K. Z i n k e . — Z. analyt. Chem. 225 (1967), 342.
71. K i v a , N.K., E.M. R o i z e n b l a t , and Kh. Z. B r a i n i n a . — In: Metody Analiza i
    Kontrolya Proizvodstva v Khimicheskoi Promyshlennosti. Scient. Res. Inst. Tech. Econ.
    Res. State Committee Council Ministers, USSR, for Chemistry, No.1. 1967.
72. B r a i n i n a , Kh.Z., T.A. R y g a i l o , and V.B. B e l y a v s k a y a . — Zavodskaya Laboratoriya
    29 (1963), 393.

73. Kolpakova, N.A., A.A. Kaplin, and A.G. Stromberg. — Zhurn. Analit. Khim. 23
(1968), 5, 665.

74. Tomoo Miwa, Synichiro Oki, and Atsushi Mizuike. — Japan Analyst 17
(1968), 7, 819.

75. Perone, S.P. and W.I. Kretlow. — Analyt. Chem. 37 (1965), 968.

76. Kolpakova, N.A., G.M. Nemtinova, and A.A. Kaplin. — Zavodskaya Laboratoriya
35 (1969), 5,529.

77. Vasil'eva, L.N. and Z.L. Yustus. — Elektrokhimiya 3 (1967), 953.

78. Roizenblat, E.M. and Kh. Z. Brainina. — Elektrokhimiya 5 (1969), 4, 396.

79. Brainina, Kh.Z., E.Ya. Neiman, and G.M. Dolgopolova. — Zavodskaya
Laboratoriya 36 (1970), 7.

80. Kiva, N.K. and Kh. Z. Brainina. — In: Sbornik Trudov Sverdlovskogo Instituta Narodnogo
Khozyaistva, 1970.

81. Brainina, Kh.Z. and N.K. Kiva. — In: Metody Analiza Khimicheskikh Reaktivov i
Preparatov, No.5—6 (1963), 134.

82. Perone, S.P. and H.E. Stapelfeldt. — Analyt. Chem. 38 (1966), 796.

83. Brainina, Kh.Z. — Doklady Akad. Nauk SSSR 130 (1960), 4, 797.

84. Roizenblat, E.M., V.B. Belyavskaya, T.A. Krapivkina, and Kh.Z. Brainina. —
In: Metody Analiza i Kontrolya Proizvodstva v Khimicheskoi Promyshlennosti, Scient.
Inst. Tech. Econ. Res. State Committee Council Ministers, USSR, for Chemistry, No.2,
1967.

85. Roizenblat, E.M., Kh.Z. Brainina, and E.Ya. Sapozhnikova. — In: Materialy
4-go Vsesoyuznogo Soveshchaniya po Polyarografii, p.150.   Alma-Ata, 1969.

86. Tindall, G.W. and S. Bruckenstein. — J. Electroanal. Chem. 22 (1969), 3, 367.

87. Vol, A.E. Structure and Properties of Binary Metallic Systems, Vol.I. — Fizmatgiz, 1959.
(Russian)

88. Khansen, Z.M. and K. Anderko. Binary Alloy Structures, Vols. 1 and 2. —
Metallurgizdat, 1962. (Russian)

89. Fedot'ev, N.P., N.N. Bibikov, P.M. Vyacheslavov, and S.Ya. Grilikhes.
Electrolytic Alloys. — Mashgiz, 1962. (Russian)

90. Igolinskii, V.A. Thesis. — Tomsk, 1964. (Russian)

91. Brainina, Kh. Z. — Doklady Akad. Nauk SSSR 154 (1964), 665.

92. Brainina, Kh. Z. — In: Trudy Komissii po Analiticheskoi Khimii 15 (1965), 185.

93. Stromberg, A.G. — In: Izv. Sib. Otd. Akad. Nauk, SSSR (1962), 76.

94. Ferrel, T.Jr. and W.C. Vosburgh. — J. Electrochem. Soc. 98 (1951), 334.

95. Huber, C.O. and L. Lemmert. — Analyt. Chem. 38 (1966), 128.

96. Brainina, Kh.Z. and N.K. Kiva. — Zhurn. Analit. Khim. 22 (1967), 4, 536.

97. Brainina, Kh.Z. — Zhurn. Analit. Khim. 19 (1964), 810.

98. Brainina, Kh.Z. and T.A. Rygailo. — Zavodskaya Laboratoriya 31   (1965), 1,28.

99. Krapivkina, T.A. and Kh.Z. Brainina. — Zavodskaya Laboratoriya 33 (1967), 400.

100. Brainina, Kh.Z. and N.K. Kiva. — Zhurn. Analit. Khim. 20 (1965), 1306.

101. Brainina, Kh.Z. and E.M. Roizenblat. — In: Materialy 2-go Vsesoyuznogo Soveshchaniya
po Polyarografii, p.16. Kazan, 1962.

102. Brainina, Kh.Z. — Zhurn. Analit. Khim. — 21 (1966), 529.

103. Brainina, Kh.Z. and T.A. Krapivkina. — Zhurn. Analit. Khim. 22 (1967), 74.

104. Brainina, Kh.Z. and T.A. Krapivkina. — Zhurn. Analit. Khim. 22 (1967), 1382.

105. Brainina, Kh.Z. and E. Ya. Sapozhnikova. — Zhurn. Analit. Khim. 21 (1966), 807.

106. Brainina, Kh.Z. and E. Ya. Sapozhnikova. — Zhurn. Analit. Khim. 21 (1966), 1342.

107. Brainina, Kh.Z. and E. Ya. Sapozhnikova. — Zhurn. Analit. Khim. 23 (1968), 231.

108. Brainina, Kh.Z. and E.M. Roizenblat. — Zhurn. Analit. Khim. 18 (1963), 1362.

109. Skyanavi-Grigor'eva, M.S. and V.I. Staroverova. — Zhurn. Org. Khim. **28** (1958), 133.

110. Lipchinskii, A.L. — Zhurn. Analit. Khim. 13 (1958), 402.

111. Tsfasman, S.B. and R.M. Salikhdzhanova. — Zavodskaya Laboratoriya 30 (1964), 133.

112. Hrabánková, E., J. Doležal, and V. Mašín. — J. Electroanal. Chem. 22 (1969), 2, 195.

113. Vetter, K. and H. Egger. — In: Materialy 14-go Soveshchaniya TsITTsE, 1963.

114. Kornfeil, F. — J. Electrochem. Soc. 109 (1962), 349.

115. Vosburgh, W.C. and Lou Pao-soong. — J. Electrochem. Soc. 108 (1961), 485.

116. Ferrel, T. Jr. and W.C. Vosburgh. — J. Electrochem. Soc. 98 (1951), 334.

117. Drotschman, C. — Chem. Abstr. 58 (1963), 226c, 8631c, 5255h.

118. Johnson, R.S. and W.C. Vosburgh. — J. Electrochem. Soc. 100 (1953), 471.

119. Coleman, J.J. — Trans. Electrochem. Soc. 90 (1946), 545.

120. Vosburgh, W.C. — J. Electrochem. Soc. 106 (1959), 839.

121. Rabano, Zh., B. Morin'ya, and Zh. Loran. — In: Materialy 14-go Soveshchaniya TsITTsE. Moscow, 1963.

122. Hrabánková, E., J. Doležal, and P. Beran. — J. Electroanal. Chem. 22 (1969), 2, 203.

123. Rutberg, L.G. Author's summary of dissertation. — Moscow, 1968.

124. Demkin, A.M. and C.I. Sinyakova. — In: Izv. AN SSSR, Seriya khimicheskaya **7** (1968), 1620.

125. Kuznetsov, V.I. — Uspekhi Khimii 18 (1949), 75.

126. Brainina, Kh.Z., A.V. Chernycheva, R.P. Lesunova, and S.A. Lomonosov. — In: Materialy 4-go Vsesoyuznogo Soveshchaniya po Polyarografii, p.103. Alma-Ata, 19

127. Sapozhnikova, E.Ya. and Kh. Z. Brainina. — In: Materialy 4-go Vsesoyuznogo Soveshchaniya po Polyarografii, p.153. Alma-Ata, 1969.

128. Kolthoff, I.M. Volumetric Analysis, Vol.2. — New York, Interscience, 1947.

129. Savostina, V.M. and V.M. Peshkova. — In: Trudy Komissii po Analiticheskoi Khimii 14 (1963), 298.

130. Krapivkina, T.A. and Kh.Z. Brainina. — Zavodskaya Laboratoriya 36 (1970), 3.

131. Brainina, Kh.Z. and T.A. Krapivkina. — Anal. Letters 2, No.5 (1968), 269.

132. Podchainova, V.N. — In: Trudy Komissii po Analiticheskoi Khimii 11 (1960), 146.

133. Bode, K. — Z. analyt. Chem. 142 (1954), 6, 414; 143 (1954), 182.

134. Usatenko, Yu.I. and F.M. Tulyupa. Izv. vyssh. zavedenii, "Khimiya i Khimicheskaya Tekhnologiya" 3 (1958), 56.

135. Usatenko, Yu.I. and F.M. Tulyupa. — Zavodskaya Laboratoriya **24** (1958), 1327.

136. Usatenko, Yu.I. and F.M. Tulyupa. — Zavodskaya Laboratoriya **25** (1959), 280.

137. Usatenko, Yu.I. and F.M. Tulyupa. — Zhurn. Neorg. Khim. 4 (1959), 2495.

138. Devis, D. and E. Boudreaux. — J. Electroanal. Chem. 8 (1964), 434.

139. Boreiko, M. Author's summary of dissertation. — Sverdlovsk, 1965.

140. Jansen, A. and B. Nygaard. — Acta chem. scand. 3 (1949), 481, 497.

141. Selbin, J. and J.H. Junkin. — J. Am. Chem. Soc. 82 (1960), 1657.

142. Both, E. and J.D.H. Strickland. — J. Am. Chem. Soc. 75 (1953), 3017.

143. Okáč, A. and M. Širnek. — Colln. Czech. Chem. Commun. 24 (1959), 2699.

144. Cornish, D.C., H.P. Dibbs, F.S. Feates, D.J.G. Ives, and R.W. Pittman. — J. Am. Chem. Soc. 84 (1962), 4104.

145. Brainina, Kh.Z. and E.M. Roizenblat. — Zavodskaya Laboratoriya 27 (1961), 1, 21.

146. Roizenblat, E.M. and Kh. Z. Brainina. — Zhurn. Analit. Khim. 19 (1964), 6, 681.

147. Kemula, W., Z. Kublik, and J. Taraszewska. — Chemist. Analyst 8 (1963), 171.

148. Ball, R.G., D.L. Manning, and O. Menis. — Analyt. Chem. 32 (1960) 6, 621.

149. Ezhov, V.D. and S.I. Zhdanov. — Zavodskaya Laboratoriya 18 (1962), 9, 1058.

150. Roizenblat, E.M., T.I. Fomicheva, and Kh.Z. Brainina. — Zavodskaya Laboratoriya 32 (1966), 6, 657.

151. Quartermain, P.G. and A.G. Hill. — Analyst 85 (1960), 211.

152. Roizenblat, E.M., E.Ya. Sapozhnikova, and Kh.Z. Brainina. — In: Metody Analiza i Kontrolya Proizvodstva v Khimicheskoi Promyshlennosti, No.1. Scient. Res. Inst. Tech. Econ. Res. State Committee Council Ministers, USSR, for Chemistry, 1967.

153. Roizenblat, E.M., T.I. Fomicheva, and Kh.Z. Brainina. — In: Khimicheskie Reaktivy i Preparaty. Trudy IREA* 28 (1966), 104.

154. Shain, J. and S. Perone. — Analyt. Chem. 33 (1961), 325.

155. Roizenblat, E.M. and Kh.Z. Brainina. — In: Khimicheskie Reaktivy i Preparaty. Trudy IREA 28 (1966), 114.

156. Berge, H. and P. Jeroschewski. — Z. analyt. Chem. 207 (1965), 110.

157. Roizenblat, E.M. and Kh.Z. Brainina. — In: Khimicheskie Reaktivy i Preparaty. Trudy IREA 28 (1966), 110.

158. Berge, H. and P. Jeroschewski. — Z. analyt. Chem. 212 (1965), 278.

159. Mine, T. — J. Electrochem. Soc. Japan 31 (1963), 22.

160. Barikov, V.G. and O.A. Songina. — Zavodskaya Laboratoriya 30 (1964), 1184.

161. Pahler, C. — J. Electroanal. Chem. 14 (1967), 329.

162. Schmidt, E. and H.R. Gygax. — J. Electroanal. Chem. 13 (1967), 378.

163. Jacquot, D. and J. Bye. — Bull. Soc. chim. Fr. 4 (1967), 1273.

164. Demkin, A.M. and S.I. Sinyakova. — In: Izv. Akad. Nauk SSSR, Seriya Khimicheskaya, No.7. (1968), 1620.

165. Adams, R. Progress in Polarography, Vol.2, p.503. — Pergamon Press, 1962.

166. Frumkin, A.N. — Uspekhi Khimii 18 (1949), 9.

167. Gailor, V.F., P.I. Elving, and A.L. Conrad. — Analyt. Chem., 25 (1953), 1620.

168. Lord, S.S. and L.B. Rogers. — Analyt. Chem. 26 (1954), 286.

169. Gailor, V.F., A.L. Conrad, and G.H. Landerl. — Analyt. Chem. 29 (1957), 224.

170. Brainina, Kh.Z. and N.K. Kiva. — Zavodskaya Laboratoriya 29 (1963), 526.

171. Roizenblat, E.M. and Kh. Z. Brainina. — Zavodskaya Laboratoriya 32 (1966), 12, 1450.

172. Brainina, Kh.Z., E.M. Roizenblat, V.B. Belyavskaya, N.K. Kiva, and T.I. Fomicheva. — Zavodskaya Laboratoriya 33 (1967), 274.

173. Brainina, Kh.Z. and V.B. Belyavskaya. — Zavodskaya Laboratoriya 31 (1965), 1172.

174. Perone, S.P. — Analyt. Chem. 35 (1963), 2091.

175. Miller, F.J. and H.E. Zittel. — J. Electroanal. Chem. 7 (1964), 116.

176. Miller, F.J. and H.E. Zittel. — Analyt. Chem. 35 (1963), 1866.

177. Beilby, A.L., W. Brooks, and G.L. Lawrence. — Analyt. Chem. 36 (1964), 22.

178. Takayoshi, Ycushimori, Masao Arakawa, and Taugio Takenchi. — Talanta 12 (1965), 147.

179. Zittel, H.E. and F.J. Miller. — Analyt. Chem. 37 (1965), 200.

180. Vassos, B.H. and H.B. Mark Jr. — J. Electroanal. Chem. 13 (1967), 1.

181. Adams, R.N. — Analyt. Chem. 30 (1958), 1576.

182. Mueller, Th.R., C.L. Olson, and R.N. Adams. Advances in Polarography, Vol.1, p.198. — Pergamon Press, 1960.

183. Olson, C.L. and R.N. Adams. — Analytica chim. Acta 22 (1960), 582.

184. Olson, C.L. and R.N. Adams. — Analytica chim. Acta 29 (1963), 358.

* [Vsesoyuznyi Nauchno- Issledovatel'skii Institut Khimicheskikh Reaktivov i Osobo Chystykh Khimicheskikh Veshchestv (All-Union Scientific Research Institute of Chemical Reagents and Ultrapure Chemical Substances).]

185. Jacobs, E.S. — Analyt. Chem. 35 (1963), 2112.

186. Farsang, G. and L. Tomcsani. — Acta Chim. Hung. 52 (1967), 2, 123.

187. Pungor, E. and E. Szepesváry. — Analytica chim. Acta 43 (1968), 289.

188. Covington, J.R. and R.J. Lacoste. — Analyt. Chem. 37 (1965), 420.

189. Igolinskii, V.A. — In: Metody Analiza Khimicheskikh Reaktivov i Preparatov, Nos.5—6 p.29. IREA, 1963.

190. Moros, S.A. — Analyt. Chem. 34 (1962), 1584.

191. Ramaley, L., R.L. Brumbaker, and C.G. Enke. — Analyt. Chem. 35 (1963), 1088.

192. Roe, D.K. and J.E.A. Toni. — Analyt. Chem. 37 (1965), 1503.

193. Matson, W.R., D.K. Roe, and D.E. Carriett. — Analyt. Chem. 37 (1965), 1594,

194. Perone, S.P. and K.K. Davenport. — J. Electroanal. Chem. 12 (1966), 269.

195. Perone, S.P. and A. Brumfield. — J. Electroanal. Chem. 13 (1967), 124.

196. Chulkina, L.S., S.I. Sinyakova, and E.K. Vul'fson. — Zhurn. Analit. Khim. (1970).

197. Crisdale, R.O., A.C. Pfister, and W. van Roosbroeck. — Bell Syst. Tech. J.30 (1951), 271.

198. Brainina, Kh.Z. — Dokl. Akad. Nauk SSSR 130 (1960), 4, 797.

199. Kaplan, B. Ya. and A.S. Rezakova. — Zhurn. Analit. Khim. 21 (1966), 10, 1268.

200. Barendecht, E. Electroanalytical Chemistry, Vol.2, p.53. — New York, 1967.

201. Barikov, V.G., Z.B. Rozhdestvenskaya, and O.A. Songina. — Zavodskaya Laboratoriya 35 (1969), 7, 776.

202. Brainina, Kh.Z. and V.B. Belyavskaya. — Ukr. Khim. Zh., No.4 (1965), 398.

203. Brainina, Kh.Z. and E.M. Roizenblat. — In: Materialy 4-go Vsesoyuznogo Soveshchaniya po Polyarografii, p.101. Alma-Ata, 1969.

204. Tur'yan, Ya. and A.M. Murenkov. — Zavodskaya Laboratoriya 27 (1961), 507.

205. Kaplan, B.Ya. and I.A. Sorokovskaya. — Zavodskaya Laboratoriya 30 (1964), 1177.

206. Neiman, E. Ya., G.M. Dolgopolova, and L.M. Trukhacheva. — Zavodskaya Laboratoriya 35 (1969), 1040.

207. Pats, R.G. and T.V. Semochkina. — Zavodskaya Laboratoriya 33 (1967), 12, 1491.

208. Pats, R.G., L.N. Vasil'eva, and T.V. Semochkina. — Zhurn. Analit. Khim. 23 (1968), 2, 241.

209. Ruby, W.C. and C.G. Tremmel. — J. Electroanal. Chem. 18 (1968), 3, 231.

210. Songina, O.A., Z.B. Rozhdestvenskaya, and V.G. Barikov. — In: Materialy 4-go Vsesoyuznogo Soveshchaniya po Polyarografii, p.90. Alma-Ata, 1969.

211. Berge, H. and P. Jeroschewski. — Z. analyt. Chem. 230 (1967), 259.

212. Berge, H. and P. Jeroschewski. — Z. analyt. Chem. 228 (1967), 9.

213. Kolpakova, N. and A.I. Kartushinskaya. — In: 5-oe Vsesoyuznoe Soveshchanie po Polyarografii (tezisy dokladov), p.79, Alma-Ata, 1969.

214. Brainina, Kh.Z. and E. Ya. Neiman. — In: 4-oe Vsesoyuznoe Soveshchanie po Polyarografii (tezisy dokladov), p.100. Alma-Ata, 1969.

215. Savostina, V.M. and V.M. Peshkova. — In: Trudy Komissii po Analiticheskoi Khimii 14 (1963), 289.

216. Gedunov, S.K. and V.S. Ryaben'kii. Introduction to the Theory of Difference Schemes. 1962. (Russian)

217. Berezin, B.S. and N.P. Zhidkov. Calculation Methods, Vol.2. — Fizmatgiz, 1959. (Russian)

218. Ditkin, V.A. and P.I. Kuznetsov. Guide to Operational Calculus, p.133. 1951. (Russian)

219. Watson, G. A Treatise on the Theory of Bessel Functions, 2nd ed. — Cambridge Univ. Press, 1966.

220. Nicholson, M.M. — Analyt. Chem. 32 (1960), 1058.
221. Perone, S.P. and J.R. Birk. — Analyt. Chem. 37 (1965), 9.
222. Neiman, E.Ya., G.M. Dolgopolova, and L.M. Trukhacheva. — Zavodskaya
     Laboratoriya 36 (1970), 636.
223. Neiman, E. Ya. and G.M. Dolgopolova. Electrochemistry and Corrosion. — In:
     Sbornik Trudov GIPROTsMO*, No.31, p.123. "Metallurgiya," 1970.
224. Brainina, Kh. Z., E. Ya. Neiman, and G.M. Dolgopolova. — Zavodskaya
     Laboratoriya 36 (1970), 783.

* [Gosudarstvennyi Nauchno-Issledovatel'skii i Proektnyi Institut Splavov i Obrabotki Tsvetnykh
  Metallov (State Scientific Research and Planning Institute of Alloys and the Working of Non-
  ferrous Metals).]